HYDROCARBON

Edited by **Vladimir Kutcherov**
and **Anton Kolesnikov**

Hydrocarbon
http://dx.doi.org/10.5772/2722
Edited by Vladimir Kutcherov and Anton Kolesnikov

Contributors

Vladimir G. Kutcherov, Ryoichi Nakada, Yoshio Takahashi, Snežana Maletić, Božo Dalmacija, Srđan Rončević, Arezoo Dadrasnia, N. Shahsavari, C. U. Emenike, Daniela M. Pampanin, Magne O. Sydnes, Diana V. Cortés-Espinosa, Ángel E. Absalón, Aizhong Ding, Yujiao Sun, Junfeng Dou, Lirong Cheng, Lin Jiang, Dan Zhang, Xiaohui Zhao, Kerstin E. Scherr, M.A. Rasheed, D.J. Patil, A.M. Dayal

Published by InTech

Janeza Trdine 9, 51000 Rijeka, Croatia

Publishing Process Manager Oliver Kurelic
Technical Editor InTech Prepress, Novi Sad
Cover InTech Design team

Additional hard copies can be obtained from orders@intechopen.com

Hydrocarbon, Edited by Vladimir Kutcherov and Anton Kolesnikov
p. cm.
ISBN 978-953-51-0927-3

For EU product safety concerns:
IN TECH d.o.o., Prolaz Marije Krucifikse Kozulić 3, 51000 Rijeka, Croatia,
info@intechopen.com or visit our website at intechopen.com.

Meet the editors

Vladimir G. Kutcherov graduated from the Gubkin Russian State University of Oil and Gas (Moscow) as a petroleum engineer in 1977. He is currently an Adjunct Professor at the Department of Energy Technology at the Royal Institute of Technology (Stockholm), and a Professor at the Department of Physics at Gubkin Russian State University of Oil and Gas (Moscow). His main research interests deal with the problem of the origin of hydrocarbons and oil and gas deposits accumulation. Prof. Kutcherov and his international collaborators have published a set of papers in PNAS, Nature Geoscience, Review of Geophysics, where the possibility of abiogenic synthesis of complex hydrocarbon systems was convincingly demonstrated.

Dr. Anton Kolesnikov was born on September 16th,1983, in Moscow, Russia, where he still lives. He is happily married, and a father of two daughters. He holds a Ph.D. in Physical Chemistry, MSAFCT, Moscow, Russia, and a Ph.D. in Energy Technology, KTH, Stockholm, Sweden. Dr. Kolesnikov is a senior researcher of the "Investigations of hydrocarbon systems in wide range of thermobaric conditions" laboratory, a senior teacher at the Physics department, and a lecturer of the "Philosophy and methodology of Science" at Gubkin Russian State University of Oil and Gas, Moscow, Russia. His research interests consist of hydrocarbons, high pressure research, theories of petroleum formation, Passivhaus concept, permaculture, philosophy and psychology of consciousness.

Contents

Chapter 1 **Abiogenic Deep Origin of Hydrocarbons and Oil and Gas Deposits Formation 1**
Vladimir G. Kutcherov

Chapter 2 **Thermogenic Methane with Secondary Alteration in Gases Released from Terrestrial Mud Volcanoes 29**
Ryoichi Nakada and Yoshio Takahashi

Chapter 3 **Petroleum Hydrocarbon Biodegradability in Soil – Implications for Bioremediation 43**
Snežana Maletić, Božo Dalmacija and Srđan Rončević

Chapter 4 **Remediation of Contaminated Sites 65**
Arezoo Dadrasnia, N. Shahsavari and C. U. Emenike

Chapter 5 **Polycyclic Aromatic Hydrocarbons a Constituent of Petroleum: Presence and Influence in the Aquatic Environment 83**
Daniela M. Pampanin and Magne O. Sydnes

Chapter 6 **Phenanthrene Removal from Soil by a Strain of *Aspergillus niger* Producing Manganese Peroxidase of *Phanerochaete chrysosporium* 119**
Diana V. Cortés-Espinosa and Ángel E. Absalón

Chapter 7 **Characterizing Microbial Activity and Diversity of Hydrocarbon-Contaminated Sites 137**
Aizhong Ding, Yujiao Sun, Junfeng Dou, Lirong Cheng, Lin Jiang, Dan Zhang and Xiaohui Zhao

Chapter 8 **Extracellular Electron Transfer in *in situ* Petroleum Hydrocarbon Bioremediation 161**
Kerstin E. Scherr

Chapter 9 **Microbial Techniques for Hydrocarbon Exploration 195**
M.A. Rasheed, D.J. Patil and A.M. Dayal

Chapter 1

Abiogenic Deep Origin of Hydrocarbons and Oil and Gas Deposits Formation

Vladimir G. Kutcherov

Additional information is available at the end of the chapter

http://dx.doi.org/10.5772/51549

1. Introduction

The theory of the abiogenic deep origin of hydrocarbons recognizes that the petroleum is a primordial material of deep origin [*Kutcherov, Krayushkin 2010*]. This theory explains that hydrocarbon compounds generate in the asthenosphere of the Earth and migrate through the deep faults into the crust of the Earth. There they form oil and gas deposits in any kind of rock in any kind of the structural position (Fig. 1). Thus the accumulation of oil and gas is

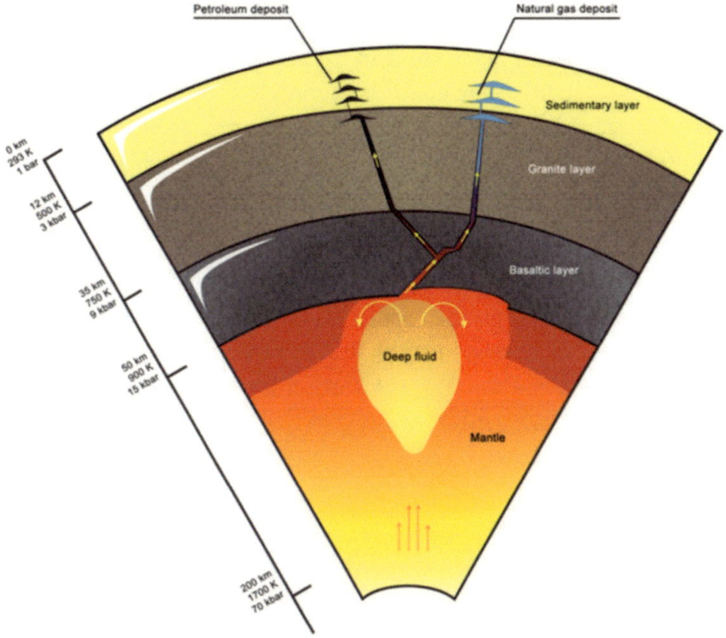

Figure 1. A scheme of genesys of hydrocarbons and petroleum deposits formation.

considered as a part of the natural process of the Earth's outgrassing, which was in turn responsible for creation of its hydrosphere, atmosphere and biosphere. Until recently the obstacles to accept the theory of the abyssal abiogenic origin of hydrocarbons was the lack of the reliable and reproducible experimental results confirming the possibility of the synthesis of complex hydrocarbon systems under the conditions of the asthenosphere of the Earth.

2. Milestones of the theory of abiogenic deep origin of hydrocarbons

According to the theory of the abyssal abiogenic origin of hydrocarbons the following conditions are necessary for the synthesis of hydrocarbons:

- adequately high pressure and temperature;
- donors/sources of carbon and hydrogen;
- a thermodynamically favorable reaction environment.

Theoretical calculations based on methods of modern statistical thermodynamics have established that:

- polymerization of hydrocarbons takes place in the temperature range 600-1500 degrees C and at pressures range of 20-70 kbar [*Kenney et al.*, 2002];
- these conditions prevail deep in the Earth at depths of 70-250 km [*Carlson et al.* 2005].

Thermobaric conditions

The asthenosphere is the layer of the Earth between 80-200 km below the surface. In the asthenosphere the temperature is still relatively high but the pressure is greatly reduced comparably with the low mantle. This creates a situation where the mantle is partially melted. The asthenosphere is a plastic solid in that it flows over time. If hydrocarbon fluids could be generated in the mantle they could be generated in the asthenosphere zone only. In the paper [*Green et al.* 2010] published in *Nature* the modern considerations about thermobaric conditions on the depth down to 200 km are shown (Fig. 2).

Composition

Donors of carbon

Mao et al., 2011 demonstrate that the addition of minor amounts of iron can stabilize dolomite carbonate in a series of polymorphs that are stable in the pressure and temperature conditions of subducting slabs, thereby providing a mechanism to carry carbonate into the deep mantle. In [*Hazen et al.*, 2012] authors suggest that deep interior may contain more than 90% of Earth's carbon. Possible sources of the carbon in the crust are shown on the Fig. 3.

Donors of hydrogen

Experimental data published in *Nature* recently [*Green et al.* 2010] shows that water-storage capacity in the uppermost mantle "is dominated by pargasite and has a maximum of about 0.6 wt% H_2O (30% pargasite) at about 1.5 GPa, decreasing to about 0.2 wt% H2O (10% pargasite) at 2.5 GPa". Another possible source of hydrogen is hydroxyl group in some minerals (biotite, muscovite).

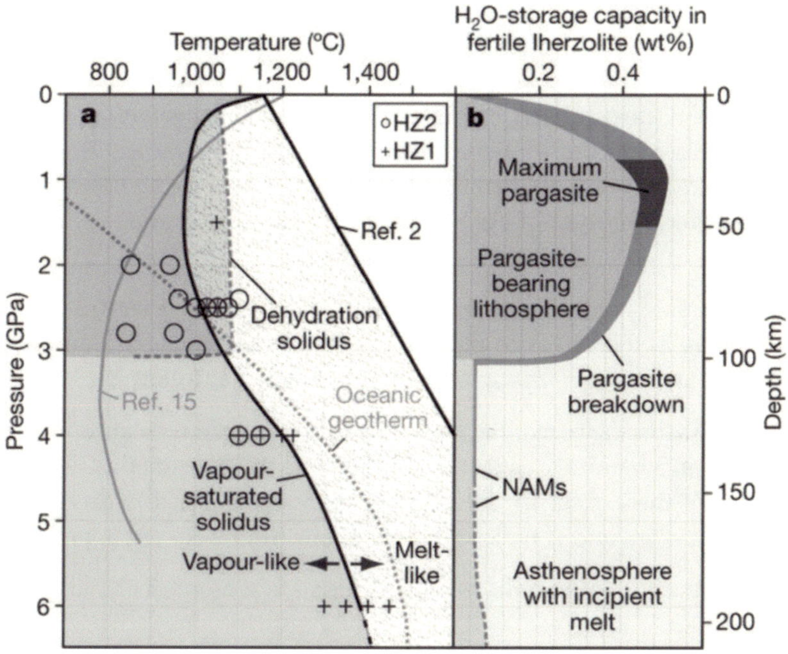

Figure 2. Thermobaric conditions in the depth [*Green et al.* 2010].

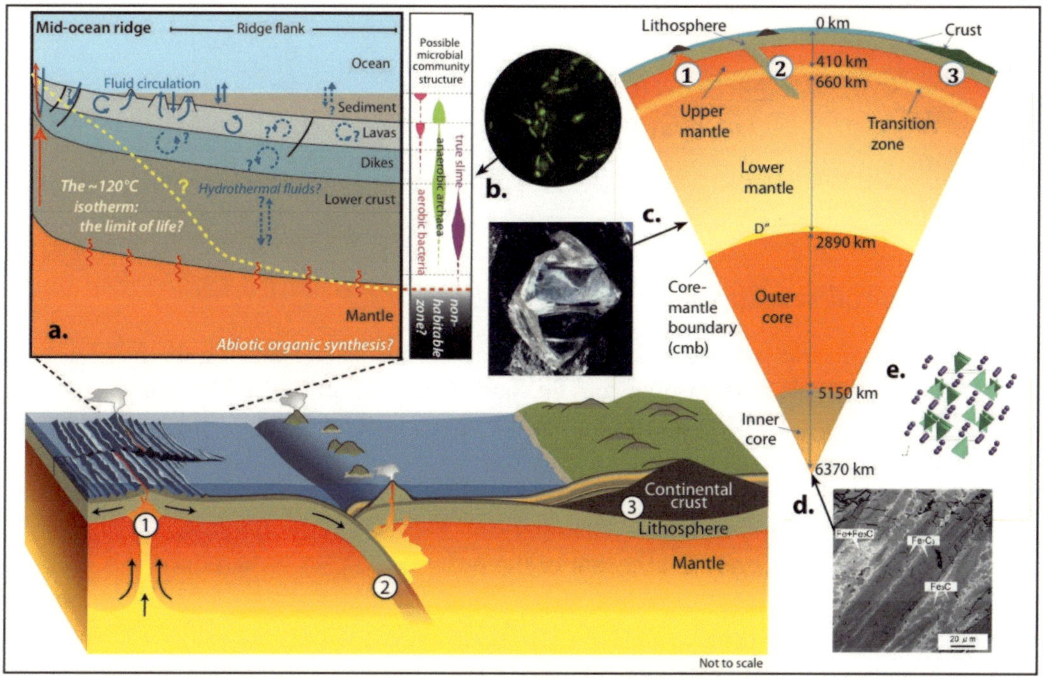

Figure 3. Sources of the carbon in the crust (adapted from images by R. Coggon and K. Nakamura).

The conclusion from the presented data is the following.

On the depth of 100 km temperature is about 1250 K and pressure is 3 GPa. On the depth of 150 km temperature is about 1500-1700 K and pressure is 5 GPa.

Both donors of carbon (carbon itself, carbonates, CO_2) and hydrogen (water, hydroxyl group of minerals) are present in the asphenosphere in sufficient amounts. Thermodynamically favorable reaction environment (reducing conditions) could be created by a presence of FeO. The presence of several present of FeO in basic and ultra-basic rocks of asthenosphere is documented.

Thus, abiogenic synthesis of hydrocarbons can take place in the basic and ultra-basic rocks of the asthenosphere in the presence of FeO, donors/sources of carbon and hydrogen. The possible reaction of synthesis in this case could be presented as follows:

- reduced mantle substance + mantle gases → oxidized mantle substance + hydrocarbons or
- combination of chemical radicals (methylene (CH_2), methyl (CH_3). Different combinations of these radicals define all scale of oil-and-gas hydrocarbons, and also cause close properties and genetic similarity of oils from different deposits of the world.

Major element composition of various mantle materials is presented in Table 1 [*Carlson et al.*, 2005].

	Fertile	*Oceanic*	*Massif*	*Off-Craton*	*Low-T Craton*	*High-T Craton*
SiO_2	45.40	44.66	44.98	44.47	44.18	44.51
TiO_2	0.22	0.01	0.08	0.09	0.02	0.11
Al_2O_3	4.49	0.98	2.72	2.50	1.04	0.84
FeO	8.10	8.28	8.02	8.19	6.72	8.08
MgO	36.77	45.13	41.15	41.63	46.12	44.76
CaO	3.65	0.65	2.53	2.44	0.54	1.08

Table 1. Major element composition of various mantle materials

3. Experimental results

One of the first reliable and reproducible experimental results confirming the possibility of hydrocarbons synthesis under upper mantle conditions were published by [*Kutcherov et al.*, 2002]. The authors used a CONAC high-pressure chamber. The stainless-steel and platinum capsules with the volume of 0.6 cm^3 were used for the experiments. The filled capsule was placed into the high-pressure chamber, pressurized and then heated to certain pressure and temperature. The treatment in the chamber for the necessary time was followed by quenching to room temperature at a rate of the order of 500 K/s, whereupon the pressure was decreased to normal pressure. The composition of the reaction products was studied by mass spectroscopy, gas chromatography, and X-ray. The initial reactants in the synthesis were chemically pure FeO (wustite), chemically pure $CaCO_3$ (calcite), and double distilled water. Tentative experiments showed that the synthesis is described by the reaction:

$$nCaCO_3 + (9n+3)FeO + (2n+1)H_2O \rightarrow nCa(OH)_2 + (3n+1)Fe_3O_4 + C_nH_{2n+2} \qquad (1)$$

where n≤11.

[*Scott et al.*, 2004] used the diamond anvil cell method with laser and resistive heating to investigate in-situ the behaviour of the system $CaCO_3$-FeO-H_2O at a pressure of 5-11 GPa and for a temperature range of 800 – 1500 K. Raman spectrometry and X-ray synchrotron diffraction methods were used for the analysis. In general, the reaction scheme proposed by [*Kutcherov et al.*, 2002] was confirmed by Scott et al.'s experimental results:

$$CaCO_3 + 12FeO + H_2O \rightarrow CaO + 4Fe_3O_4 + CH_4 \qquad (2)$$

A comparison of [*Kutcherov et al.*, 2002] and [*Scott et al.*, 2004] experimental results, where the same initial reactants and thermobaric conditions were used, shows us the following. In the case of Raman spectrometry in the diamond anvil cell [*Scott et al.*, 2004] only methane was found. In [*Kutcherov et al.*, 2002] the mixture of hydrocarbons up to hexane was detected. The difference may be explained by different reasons. Quantities of saturated hydrocarbons heavier than methane, on an analogy with [*Kutcherov et al.*, 2002] should decrease dramatically with an increase in molecular weight of hydrocarbons. The sensitivity of the Raman spectrometry method in the DAC is not enough to detect the saturated hydrocarbons in comparison with relatively large volume experiments in the CONAC chamber with gas chromatography. Also fluorescence from iron-containing materials in the cell could be the obstacle to analyze these saturated hydrocarbons dispersed in the diamond anvil cell.

To check the above-mentioned explanation the behaviour of pure methane at pressure range 1-14 GPa at the temperature interval 900-2500 K was studied [*Kolesnikov et al.*, 2009]. The experiments have shown that methane is stable at temperature below 900 K in the above-mentioned pressure range. In the temperature interval 900-1500 K at 2-5 GPa the formation of heavier alkanes (ethane, propane, and butane) from methane was observed:

$$CH_4 \rightarrow C_2H_6 + C_3H_8 + C_4H_{10} + H_2 \qquad (3)$$

At temperature higher that 1500 K methane dissociates to carbon (graphite) and molecular hydrogen:

$$CH_4 \rightarrow C + H_2 \qquad (4)$$

In [*Kolesnikov et al.*, 2009] the the redox conditions of the mantle were also modeled by introducing into a system of magnetite (Fe_3O_4), which was partially transformed to iron (0) forming a redox buffer.

The introduction of Fe_3O_4 did not affect the thermobaric conditions of methane transformation. Pure iron and water were detected in the reaction products. This gives us the possibility to suggest the following pathway of reactions:

$$CH_4 + Fe_3O_4 \rightarrow C_2H_6 + Fe + H_2O \qquad (5)$$

$$CH_4 + Fe_3O_4 \rightarrow C + Fe + H_2O \quad (6)$$

The experimental results confirming the possibility of synthesis of natural gas from inorganic compounds under the upper mantle conditions were published in [*Kutcherov et al., 2010*]. The experimental results received in CONAC high-pressure chamber and in a split-sphere high-pressure device [6] are shown in Table 2. The gas chromatography and X-ray technique were used to analyze the reaction products.

If the carbon donor was $CaCO_3$ (experiments 1 and 2), the methane concentration in the produced mixture was as that in "fat" (rich in heavy hydrocarbons) natural gas (as at the Vuktilskoe gas field). If the carbon donor was individual carbon (experiments 3 and 4), the hydrocarbon composition corresponded to "dry" (methane-rich) natural gas (as at the Severo - Stavropolskoe gas field). Rapid cooling (quenching) fixes CH_4 and C_2H_6/C_2H_4 in the reaction products. After cooling for 4 h (experiment 4), the amount of CH_4 and C_2H_6/C_2H_4 in the reaction products increases by a factor of tens, and heavier hydrocarbons up to C_4H_{10}/C_4H_8 emerge. Thus, the time of cooling of the fluid forming at high pressure (e.g., in the course of its jet migration to the surface) has a significant effect on the final composition of the fluid. A decrease in the rate of cooling of the initial fluid results in synthesis of heavier saturated hydrocarbons in the mixture. Then, at the first stage, in case of quenching (experiment 3), the reaction is the following:

$$Fe + H_2O \rightarrow FeO + H_2 \quad (7)$$

and in case of slow cooling (experiment 4), the reaction is

$$3Fe + 4H_2O \rightarrow Fe_3O_4 + 4H_2 \quad (8)$$

At the second stage, in both cases, hydrocarbons can be assumed to form from carbon and hydrogen by the reactions

$$nC + (n+1)H_2 \rightarrow C_nH_{2n+2} \quad (9)$$

$$nC + nH_2 \rightarrow C_nH_{2n} \quad (10)$$

Cooling shifts the equilibrium of the reaction between Fe(0) and water toward the formation of hydrogen and stronger iron oxidation. The fact that reaction (8) involves more hydrogen than reaction (7) does results in an increase in the amounts of hydrocarbons in reactions (9) and (10). However, although the influence of the quenching time is obvious, the full explanation of this effect requires further investigation.

4. Formation of oil and gas deposits in the light of the abiogenic deep origin of hydrocarbons

The theory of the abiogenic deep origin of hydrocarbons denies the lateral migration of oil and gas in their reservours unless a hydrodynamic (hydraulic) fluid movement exists.

Experiment, reagents (mg), cooling rate	Concentration, mol.%								CH_4, μmol
	CO_2	N_2	CO	CH_4	C_2H_4 C_2H_6	C_3H_6 C_3H_8	C_4H_8 C_4H_{10}	C_5H_{10} C_5H_{12}	
Toroidal high-pressure apparatus (CONAC)									
experiment A quenching $CaCO_3$(104.5)+Fe(174.6)+ H_2O(45.3)	0.0	tr.	0,0	71.4	25.8	2.5	0.25	0	6.28
experiment B quenching $CaCO_3$(104.7)+Fe(174.6)+ D_2O(42.6)	0.0	tr.	0.0	71.1	25.3	3.2	tr.	tr.	4.83
Split-sphere high-pressure device (BARS)									
experiment C quenching C(24.5)+Fe(60.2)+H_2O(10.1)	0	0	0	96.1	3.84	0	0	0	0.23
experiment D 4 h. cooling C(21.3)+Fe(98.6)+H_2O(15.1)	0	0	0	93.2	6.21	0.42	0.16	0	5.4
Severo-Stavropolskoe gas field	0.23	-	-	98.9	0.29	0.16	0.05	-	
Vuctyl'skoe gas field	0.1	-	-	73.80	8.70	3.9	1.8	6.40	

Table 2. The experimental results received in CONAC high-pressure chamber and in a split-sphere high-pressure device

Capillary forces which are related to the pore radius and to the surface tension across the oil-water (or gas-water) interface (the process is decribed by Laplace's Equation) are, generally, 12-16 thousand times stronger than the buoyancy forces of oil and gas (according to Navier–Stockes Equation) in the natural porous, permeable media of subsurface. According to the theory of the abiogenic deep origin of hydrocarbons oil and gas accumulations are born as follows. Rising from sub-crust zones through the deep faults and their feather joints or fissures the deep petroliferous fluid is injected under high pressure into any rock and distributed there. The hydrocarbon composition of oil and gas accumulations formed this way depends on cooling rate of the fluids during their injection into the rocks of the Earth's Crust. When and where the further supply of injected hydrocarbons from the depth stops, the fluids do not move further into any forms of the Earth Crust (anticline, syncline, horizontal and tilted beds) without the re-start of the injection of the abyssal petroliferous fluids. The most convincing evidence of the above-mentioned mechanism of oil and gas deposit formations is the existence of such giant gas fields as Deep Basin (Figure 3), Milk River and San Juan (the Alberta Province of Canada and the Colorado State, U.S.A.) The formation of these giant gas fields questions the existence of any lateral migration of oil and gas during the oil and gas accumulation process. Those giant gas fields occur in synclines where gas must be generated but not be accumulated, according to the hypothesis of biotic petroleum origin and hydrodynamically controlled migration. The giant gas volumes ($12.5 \cdot 10^{12}$ cu m in Deep Basin, $935 \cdot 10^{9}$ cu m in San Juan, $255 \cdot 10^{9}$ cu m in Milk River) are concentrated in the very fine-grained, tight, impermeable argillites, clays, shales and in tight sandstones and siltstones [*Masters*, 1979]. These rocks are usually accepted as source rocks cap rocks/seals rocks in petroleum geology but by no means of universally recognized reservoir rocks of oil and natural gas. All the gas-saturated tight rocks here are graded updip into coarse-grained, highly-porous and highly-permeable aquifers with no visible tectonic, lithological and stratigraphic barriers to prevent updip gas migration. Therefore, the tremendous gas volumes of above-mentioned gas fields have the tremendous buoyancy but it never overcomes capillary resistance in pores of the water-saturated reservoir rocks. Gas is concentrated in the tight impermeable sand which is transformed progressively and continuously updip into the coarse-grained, high-porous and high-permeable sand saturated by water.

5. Natural gas and oil in the recent sea-floor spreading centers

Petroleum of abyssal abiogenic origin and their implacement into the crust of the Earth can take place in the recent sea-floor spreading centres in the oceans. Igneous rocks occupy 99 % of the total length (55000 km) of them while the thickness of sedimentary cover over the spreading centres does not exceed 450-500 m [*Rona*, 1988]. Additionally sub-bottom convectional hydrothermal systems discharge hot (170-430 degrees C) water through the sea bottom's black and white "smokers". Up to now more than 100 hydrothermal systems of this kind have been identified and studied by in scientific expeditions using submarines such as «ALWIN», «Mir», «Nautile», and «Nautilus» in the Atlantic, Pacific and Indian oceans respectively. Their observations pertaining to the deep abiogenic origin of petroleum are as follows:

- the bottom "smokers" of deepwater rift valleys vent hot water, methane, some other gases and petroleum fluids. Active "plumes" with the heights of 800-1000 m venting methane have been discovered in every 20-40 km between 12°N and 37°N along the Mid-Atlantic Ridge (MAR) over a distance of 1200 km. MAR's sites – TAG (26°N), Snake Pit (23°N), Logatchev (14°45'N), Broken Spur (29°N), Rainbow (37°17'N), and Menez Gwen (37°50'N) are the most interesting.
- At the Rainbow site, where the bottom outcrops are represented by ultramafic rocks of mantle origin the presence of the following substances were demonstrated (by chromatography/mass-spectrometry): CH_4; C_2H_6; C_3H_8; CO; CO_2; H_2; H_2S, N_2 as well as petroleum consisted of n-C_{16}–n-C_{29} alkanes together with branched alkanes, diaromatics [*Charlou et al.*, 1993, 2002]. Contemporary science does not yet know any microbe which really generates n-C_{11} – n-C_{22} alkanes, phytan, pristan, and aromatic hydrocarbons.

- At the TAG site no bottom sediments, sedimentary rocks [*Simoneit*, 1988; *Thompson et al.*, 1988], buried organic matter or any source rocks. The hydrothermal fluid is very hot (290/321 degree C) for any microbes. There are the Beggiatoa mats there but they were only found at some distances from "smokers".
- Active submarine hydrothermal systems produce the sulfide-metal ore deposits along the whole length of the East Pacific Rise (EPR). At 13°N the axis of EPR is free of any sediment but here aliphatic hydrocarbons are present in hot hydrothermal fluids of black "smokers". In the sulfide-metal ores here the methane and alkanes higher than n-C_{25} with prevalence of the odd number of carbon atoms have been identified [*Simoneit*, 1988].
- Oil accumulations have been studied by the «ALVIN» submarine and by the deep marine drilling in the Gulf of California (the Guaymas Basin) and in the Escanaba Trough, Gorda Ridge [*Gieskes et al.*, 1988; *Koski et al.*, 1988; *Kvenvolden et al.*, 1987; *Lonsdale*, 1985; *Peter et al.*, 1988; *Simoneit*, 1988; *Simoneit et al.*, 1982; *Thompson et al.*, 1988] of the Pacific Ocean. These sites are covered by sediments. However, petroleum fluids identified there are of hydrothermal origin according to *Simoneit, Lonsdale* [1982] and no source rocks were yet identified there.
- As for other sites over the Globe, scientific investigations with the submarines have established that the methane "plumes" occur over the sea bottom "smokers" or other hydrothermal systems in Red Sea, near Galapagos Isles, in Mariana and Tonga deepwater trenches, Gulf of California, etc. [*Baker et al.*, 1987; *Blanc et al.*, 1990; *Craig et al.*1987; *Evans et al.*, 1988; *Horibe et al.*, 1986; *Ramboz et al.*, 1988]. Non-biogenic methane (10^5-10^6 cu m/year) released from a submarine rift off Jamaica [*Brooks*, 1979] has been also known.

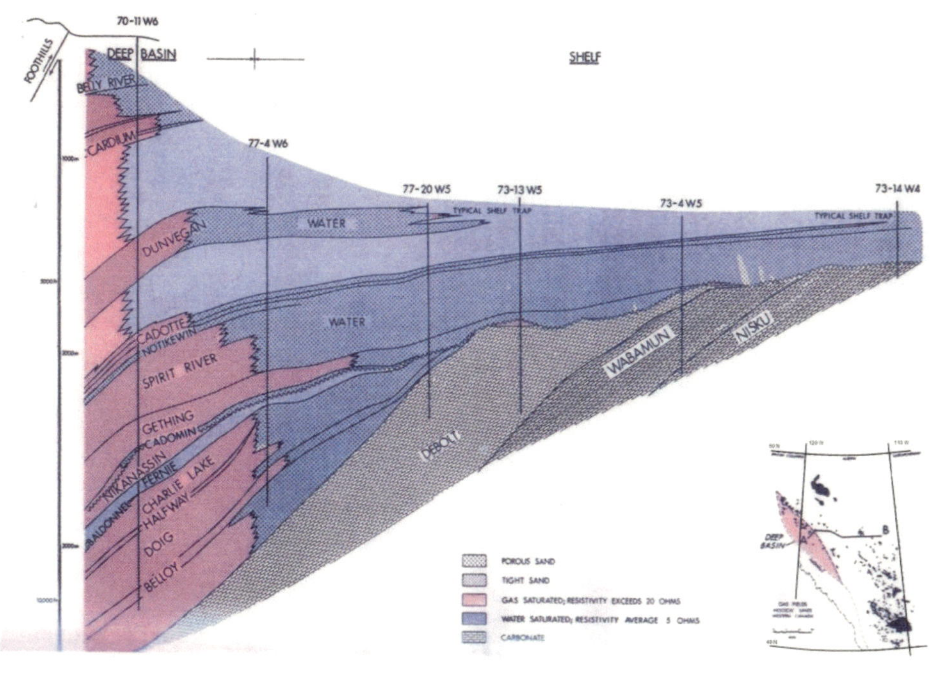

Figure 4. Cross section of Alberta showing gas-saturated sands of Deep Basin [*Masters* 1979].

A recent investigation along the Mid-Atlantic Ridge 2,300 miles east of Florida confirms that the hydrogen-rich fluids venting at the bottom of the Atlantic Ocean in the Lost City Hydrothermal Field were produced by an abiotic synthesis of hydrocarbons in the Mantle of the Earth (Fig. 3) [*Proskurowski et al.*, 2008]. Quantitatively speaking the sea-floor spreading centres may vent $1.3 \cdot 10^9$ cu m of hydrogen and $16 \cdot 10^7$ cu m methane annually [*Welhan et al.*, 1979].

Data discussed below confirm the following:

1. Source rocks accounting for the volume of the petroleum venting described are not available;
2. The natural gas and petroleum fluids in the recent sea-floor spreading centres can be explained as a result of the vertical migration of the Mantle fluids.

6. Bitumen and hydrocarbons in native diamonds

A presence of bitumen and hydrocarbons in native diamonds, carbonado and kimberlites could be taking into a consideration as the evidence confirming the abyssal petroleum origin. Studying the native diamonds, carbonado and kimberlites under the microscope many scientists from several countries have found the numerous primary fluid inclusions which have been opened due to the specific methods. Fluid contents of primary fluid inclusions have been recovered without any contamination and studied by mass-spectrometry/gas-chromatography. Results of such investigations carried-out on the samples from Africa, Asia, Europe, North and South America can be summarized as follows.

The well-known diamond-producing mines such as the Dan Carl, Finsh, Kimberley, and Roberts Victor are located in the kimberlite pipes of the **South Africa**. There the African shield is characterized by the remarkable disjunctive dislocations and non-orogenic magmatism which has produced a great number of the carbonatite and kimberlite intrusions and explosion pipes in the area around Lake Tanganyika, Lake Malawi and Lake Victoria between 70 Ma and 3000 Ma ago [*Irvine*, 1989]. These lakes are in the Great East-African Rift Valley. The Valley's margins and disjunctive edges consist of the African shield crystalline rocks. 258 samples of diamonds from this area have been investigated under the microscope [*Deines et al.*, 1989]. The investigation has shown the presence of primary fluid inclusions in all samples investigated. These samples have been disintegrated into the small particles in vacuum of about $1.3 \cdot 10^{-6}$ Pa and 200 degrees C. The gas mixture from each sample was received. Mass-spectrometric/gas-chromatographic studies of the mixtures are shown in Table 3.

The same hydrocarbons and gases mixtures were detected in natural diamonds from **Congo, Brazil** [*Melton et al.*, 1974] and **Zaire** [*Giardini et al.*, 1982] (Table 3).

The composition of the primary fluid inclusions composition has been studied by mass-spectrometer in seven native **Arkansas** diamonds. The result of investigation has confirmed the presence of the different kinds of hydrocarbon in all samples (Table 3) [*Melton et al.,* 1974].

In a **Brazil** carbonado primary fluid inclusions comprise a set of heavy hydrocarbons (Table 3). Pyrope ($Mg_3Al_2(SiO4)_3$) and olivine which were recovered from diamond crystals and kimberlites of the Mir, Ruslovoye, and Udatchnoye Eastern diamond-bearing pipes, **East Siberia, Russia** contain a number of different hydrocarbons (Table 3) [*Kulakova et al.*, 1982; *Kaminski et al.*, 1985]. According to *Makeev et al.* [2004] 9-to-27 forms of metallic films were found and studied upon the crystal faces of diamonds from **Brazil** and from the Middle Timan, Ural and Vishera diamonds in the European part of **Russia**. These films consist of aluminum, cadmium, calcium, chrome, cerium, copper, gold, iron, lanthanum, lead, magnesium, neodymium, nickel, palladium, silver, tin, titanium, ytterbium, yttrium, zinc, zirconium and of precious metals including even Au_2Pd_3. The thickness of these films is from fractions of micrometer to several micrometers. These films are the evidences for the growth of diamonds from carbon dissolved in the melt of gold and palladium [*Makeev, Ivanukh,* 2004]. The coarseness of the diamond crystals in kimberlite and lamprophyre pipes depends on the sizes of precious metal droplets in the respective zone – in the Earth's upper, transitional, and lower Mantle.

Region	*Gas mixture concentration, vol. %*
Africa Diamonds	CH_4, C_2H_4, C_3H_6, solid hydrocarbons, C_2H_5OH, Ar, CO, CO_2, H_2, O_2, H_2O, and N_2
Congo, Brazil, Zaire Diamonds	5.8 of CH_4; 0.4 of C_2H_4; 2.0 of C_3H_6; traces of C_4H_8; C_4H_{10} and solid hydrocarbons
Arkansas, U.S.A. diamonds	0.9-5.8 of CH_4; 0.0-5.2 of CH_3OH; 0.0-3.2 of C_2H_5OH; 1.2-9.4 of CO; 5.3-29.6 of CO_2; 1.5-38.9 of H_2; 2.9-76.9 of H_2O; 0.0-87.1 of N_2; 0.0-0.2 of Ar
Brazil carbonado	the homologies of naphthalene ($C_{10}H_8$), phenanthrene ($C_{14}H_{10}$), and pyrene ($C_{16}H_{10}$). Total concentration varies from 20 to 38.75 g/t
East Siberia, Russia diamonds and kumberlites	C_6H_6; $C_{12}H_{10}$; $C_{20}H_{12}$; $C_{16}H_{10}$, and other polynuclear aromatic hydrocarbons. Total concentration is about 0.136 g/t

Table 3. Results of Investigation of Gas Mixtures from Native Diamonds, Carbonado and Kumberlites

Investigation primary fluid inclusions in diamonds have shown a presence of bitumen in diamonds. The primary inclusions preserved in natural diamonds are bitumen inclusions and contain mantle hydrocarbons. This evidences that the source materials for the abyssal, natural synthesis of diamonds were the hydrocarbon fluids which have saturated the outgassing mantle and enabled mantle silicates to be reduced to native metals. The Brazil natural diamonds were sampled from the Juine kimberlite pipe field of the Mato Grosso State, Brazil. The Juine Later Cretaceous kimberlites contain five mineral associations related to the different facies and depths which are reflected in Table 4. One of the Juine diamonds sampled near Sao Luis Creek was the lower Mantle diamond and comprised the primary fluid inclusions with the lower Mantle bitumens [*Makeev et al.*, 2004].

$\delta^{13}C$ for 213 diamonds from the different pipes was analyzed. $\delta^{13}C$ is ranged from -1.88 to -16 ‰ [*Deines et al.*, 1989]. The chemical and isotope peculiarities of natural diamonds reflect the different Mantle media and environments. Diamonds with $\delta^{13}C$ from -15 to -16 ‰ come from the region at a lower depth than the natural diamonds with $\delta^{13}C$ from -5 to -6 ‰.

Conclusions:

1. No doubt, that diamonds, carbonado and kimberlites are formed at great depths.
2. Presence of the inhibited primary hydrocarbon inclusions in diamonds, carbonado and kimberlites testifies that hydrocarbon Mantle fluids were a material for synthesis of these minerals in the Mantle.
3. Presence of abiotic hydrocarbon fluids in the Mantle of the Earth is scientifically proved evidence.

7. Petroleum in meteor impact craters

Petroleum reserves in meteor impact craters posses a great potential. At the present moment there are about 170 meteor impact craters identified in all continents and in the world ocean bottom. Impact fracturing can occur to depths of 35-40 km and penetrate into the Earth's mantle. The impact fractures are the result of impacts of asteroids, bolides, comets on the Earth. When a massive cosmic object impacts the Earth surface with the velocity in the range from 15 to 70 km/s it accompanies the explosion. According to experiments devoted to mechanisms and models of cratering in the Earth media, the hyper fast impact creates temperature of 3000 degrees C and pressure of 600-900 kbar in the rocks of different compositions and generates their disintegration, pulverization, vaporization/exhalation, oxidation and hydrothermal transformation. As the result of the above-mentioned events and processes the meteorite (comet) impact transforms any non-reservoir rock into a porous and permeable reservoir rock [*Curran et al.*, 1977; *Masaitis et al.*, 1980; *Donofrio*, 1981].

Petroleum reserves were found in onshore and offshore meteor impact craters carbonate, sandstone and granite rocks over the world [*Donofrio*, 1998; *India, 2006*] (Table 4). Granites compose the crystalline basement of meteor impact craters whereas the carbonates and sandstones compose the sedimentary infill of the crater. Their producing depth is determined from 61 to 5185 m; the total production is from 4.8 to 333,879 m^3/d of oil and from 7363 m^3/d to $39.6 \cdot 10^6$ m^3/d of gas; the total proven reserves are from 15,899 m^3 to $4770 \cdot 10^6$ m^3 of oil, $48 \cdot 10^6$ m^3 of condensate and from $56.6 \cdot 10^6$ to $424.8 \cdot 10^9$ m^3 of gas [*Donofrio*, 1998].

The richest petroleum meteor impact crater Cantarell is in Mexico. Its cumulative production exceeds $1.1 \cdot 10^9$ m^3 of oil and $83 \cdot 10^9$ m^3 of gas. The current remnant recoverable reserves are equal to $1.6 \cdot 10^9$ m^3 of oil and $146 \cdot 10^9$ m^3 of gas in three productive zones. They produce currently 206,687 m^3/d of oil, and 70 % of it is recovered from carbonate breccia only. Its porosity is 8-12 % and the permeability is 3000-5000 millidarcies. Occurring at the

Impact crater	*Country*	*Age, Ma*	*Diameter, km*
Akreman (onshore)	Australia	600	100
Ames (onshore) oil&gas deposits in carbonates and granites	U.S.A.	470±30	15
Avak (onshore) gas deposits in sandstones	U.S.A.	3-95	12
Can-Am (onshore)	Canada and U.S.A.	500	100
Calvin (onshore) oil deposits in carbonates	U.S.A.	450±10	8.5
Chicxulub (offshore) oil&gas deposits in carbonates	Mexico	65	240
Kara (onshore)	Russia	60	65
Marquez (onshore) oil&gas deposits in carbonates and sandstones	U.S.A.	58±2	12.7
Newporte (onshore) oil&gas deposits in carbonates and sandstones	U.S.A.	<500	3.2
Popigai (onshore)	Russia	39	100
Red Wing Creek (onshore) oil&gas deposits in carbonates	U.S.A.	200±25	9
Sierra Madera (onshore) gas deposits in carbonates	U.S.A.	<100	13
Shiva (offshore)	India	15	500
South Caribbean crater (offshore)	Colombia	65	300
South Cuban crater (offshore)	Cuba	65	225
Steen River (onshore) oil deposits in carbonates and granites	Canada	91±7	25
Viewfield (onshore) oil&gas deposits in carbonates	Canada	190±20	2.5
Wredefort (onshore)	South Africa	1970	140

Table 4. Parameters of the biggest and petroleum productive impact craters

Tertiary Cretaceous boundary this breccia is genetically related to Chicxulub impact crater the diameter of which is now measured to be 240 km [*Grajales-Nishimura et al.,* 2000].

Calculating with an average porosity, permeability and water saturation of the over crater breccia and fracturity of the undercrater crystalline Earth Crust together with the rocks which arround the crater the petroleum potential of the single meteor impact crater having the diameter of 20 km can exceed the total proven oil-and-gas reserves of the Middle East [*Donofrio,* 1981]. *Donofrio* [1981] also estimates that during the last 3000 Ma the meteorite-comet bombardment of the Earth must have created 3060 onshore meteor impact craters of similar diameters. *Krayushkin* [2000] calculates with 7140 submarine meteor impact craters can be equal to about $12 \cdot 10^{14}$ m^3 of oil and $7.4 \cdot 10^{14}$ m^3 of gas.

The oil and gas in the meteor impact craters cannot be biogenic according since:

1. Any intercrater source rocks are destroyed, disintegrated, melted and pulverized together with all the other rocks at the site of the meteorite impact [*Masaitis et al.*, 1980];
2. After the impact any lateral petroleum migration from the non-crater zones into the crater through the concentric ring uplifts of 100-300 m high and concentric ring trenches of 100-300 m deep which surround the central uplift of the crater is not enabled.

8. Oil and gas deposits in the Precambrian crystalline basement

Crystalline crust of the Earth is the basement of 60 sedimentary basins with commercial oil and gas deposits in 29 countries of the world. Additionally, there are 496 oil and gas fields in which commercial reserves occur partly or entirely in the crystalline rocks of that basement. 55 of them are classified as giant fields (>500 Mbbls) with 16 non-associated gas, 9 gas-oil and 30 undersaturated oil fields among them (Table 5).

Deposit	*Country*	*Proven reserves*
Achak (gas field)	Turkmenistan	$155 \cdot 10^9$ cu m
Gugurtli (gas field)	Turkmenistan	$86 \cdot 10^9$ cu m
Brown-Bassett (gas field)	UK	$73 \cdot 10^9$ cu m
Bunsville (gas field)	U.S.A.	$85 \cdot 10^9$ cu m
Gomez (gas field)	U.S.A.	$283 \cdot 10^9$ cu m
Lockridge (gas field)	U.S.A.	$103 \cdot 10^9$ cu m
Chayandinskoye (gas field)	Russia	$1{,}240 \cdot 10^9$ cu m
Luginetskoye (gas field)	Russia	$86 \cdot 10^9$ cu m
Myldzhinskoye (gas field)	Russia	$99 \cdot 10^9$ cu m
Durian Mabok (gas field)	Indonesia	$68.5 \cdot 10^9$ cu m
Suban (gas field)	Indonesia	$71 \cdot 10^9$ cu m
Gidgealpa (gas field)	Australia	$153 \cdot 10^9$ cu m
Moomba (gas field)	Australia	$153 \cdot 10^9$ cu m
Hateiba (gas field)	Libya	$411 \cdot 10^9$ cu m
Bach Ho (oil and gas field)	Vietnam	$600 \cdot 10^6$ t of oil and $37 \cdot 10^9$ cu m of gas
Bombay High (oil and gas field)	India	$1640 \cdot 10^6$ t of oil and $177 \cdot 10^9$ cu m of gas
Bovanenkovskoye (oil and gas)	Russia	$55 \cdot 10^6$ t of oil and $2400 \cdot 10^9$ cu m of gas
Tokhomskoye (oil and gas field)	Russia	$1200 \cdot 10^6$ t of oil and $100 \cdot 10^9$ cu m of gas
Coyanosa (oil and gas field)	U.S.A.	$6 \cdot 10^6$ t of oil and $37 \cdot 10^9$ cu m of gas

Deposit	*Country*	*Proven reserves*
Hugoton-Panhandle (oil and gas)	U.S.A.	$223 \cdot 10^6$ t of oil and $2000 \cdot 10^9$ cu m of gas
Peace River (oil and gas field)	U.S.A.	$19000 \cdot 10^6$ t of oil and $147 \cdot 10^9$ cu m of gas
Puckett (oil and gas field)	U.S.A.	$87.5 \cdot 10^6$ t of oil and $93 \cdot 10^9$ cu m of gas
La Vela (oil and gas field)	Venezuela	$54 \cdot 10^6$ t of oil and $42 \cdot 10^9 \cdot 10^9$ cu m of gas
Amal (oil field)	Libya	$583 \cdot 10^6$ t
Augila-Nafoora (oil field)	Libya	$178 \cdot 10^6$
Bu Attifel (oil field)	Libya	$90 \cdot 10^6$
Dahra (oil field)	Libya	$97 \cdot 10^6$
Defa (oil field)	Libya	$85 \cdot 10^6$
Gialo (oil field)	Libya	$569 \cdot 10^6$
Intisar "A" (oil field)	Libya	$227 \cdot 10^6$
Intisar "D" (oil field)	Libya	$182 \cdot 10^6$
Nasser (oil field)	Libya	$575 \cdot 10^6$
Raguba (oil field)	Libya	$144 \cdot 10^6$
Sarir (oil field)	Libya	$1150 \cdot 10^6$
Waha (oil field)	Libya	$128 \cdot 10^6$
Claire (oil field)	UK	$635 \cdot 10^6$
Dai Hung (oil field)	Vietnam	$60\text{-}80 \cdot 10^6$
Su Tu Den (oil field)	Vietnam	$65 \cdot 10^6$
Elk Basin (oil field)	U.S.A.	$70 \cdot 10^6$
Kern River (oil field)	U.S.A.	$200.6 \cdot 10^6$
Long Beach (oil field)	U.S.A.	$121 \cdot 10^6$
Wilmington (oil field)	U.S.A.	$326 \cdot 10^6$
Karmopolis (oil field)	Brazil	$159 \cdot 10^6$
La Brea-Pariñas-Talara (oil field)	Peru	$137 \cdot 10^6$
La Paz (oil field)	Venezuela	$224 \cdot 10^6$
Mara (oil field)	Venezuela	$104.5 \cdot 10^6$
Mangala (oil field)	India	$137 \cdot 10^6$
Renqu-Huabei (oil field)	China	$160 \cdot 10^6$
Shengli (oil field)	China	$3230 \cdot 10^6$
Severo-Varieganskoye (oil field)	Russia	$70 \cdot 10^6$
Sovietsko-Sosninsko-Medvedovskoye (oil field)	Russia	$228 \cdot 10^6$

Table 5. Giant and Supergiant Petroleum Deposits in the Precambrian Crystalline Basement

They contain $9432 \cdot 10^9$ m^3 of natural gas and $32{,}837 \cdot 10^6$ tons of crude oil, i.e., 18 % of the total world proven reserves of oil and about 5.4 % of the total world proven reserves of natural gas.

In the crystalline basement the depths of the productive intervals varies of 900-5985 m. The flow rates of the wells is between 1-2 m^3/d to 2,400 m^3/d of oil and 1000-2000 m^3/d to $2.3 \cdot 10^6$ m^3/d of gas. The pay thickness in a crystalline basement is highly variable. It is 320 m in the Gomez and Puckett fields, the U.S.A.; 680 m in Xinglontai, China; 760 m in the DDB's northern flank. The petroleum saturated intervals are not necessarity right on the top of the crystalline basement. Thus, oil was discovered at distance of 18-20 m below the top of crystalline basement in La Paz and Mara fields (Western Venezuela), 140 m below the basement's top in the Kazakhstan's Oimasha oil field. In the Baltic Shield, Sweden the 1 Gravberg well produced 15 m^3 of oil from the Precambrian igneous rocks of the Siljan Ring impact crater at the depth of 6800 m. In the Kola segment of the Baltic Shield several oil-saturated layers of the Precambrian igneous rocks were penetrated by the Kola ultra-deep well at the depth range of 7004-8004 m.

One of the most success stories of the practical application of the theory of the abyssal abiogenic origin of petroleum in the exploration is the exploration in the Dnieper-Donetsk Basin (DDB), Ukraine [*Krayushkin et al.* 2002]. It is a cratonic rift basin running in a NW-SE direction between 30.6°E-40.5°E. Its northern and southern borders are traced from 50.0°N-51.8°N and 47.8°N-50.0°N, respectively. In the DDB's northern, monoclinal flank the sedimentary sequence does not contain any salt-bearing beds, salt domes, nor stratovolcanoes and no sourse rocks. Also this flank is characterized by a dense network of the numerous syn-thethic and anti-thethic faults. These faults create the mosaic fault-block structure of crystalline basement and its sedimentary cover, a large number of the fault traps (the faulted anticlines) for oil and natural gas, an alternation uplifts (horsts) and troughs (grabens). The structure of the DDB's northern flank excludes any lateral petroleum migration across it from either the Donets Foldbelt or the DDB's Dnieper Graben.

Consequently, the DDB's northern flank earlier was qualified as not perspective for petroleum production due to the absence of any "source rock of petroleum" and to the presence of an active, highly dynamic artesian aquifer. However, after a while the perspectivity of this area was re-interpreted, re-examined in compliance with the theory of the abyssal abiogenic origin of petroleum starting with the detailed analysis of the tectonic history and geological structure of the crystalline basement in the DDB's northern monoclinal flank. Subsequently respective geophysical and geochemical procpecting programmes were accepted primarily for exploring deep-seated petroleum.

Late 1980's-early 1990's 61 wells were drilled in the DDB's northern flank. 37 of them proved commercially productive (the exploration success rate is as high as 57 %) discovering commercial oil and gas strikes in the Khukhra, Chernetchina, Yuliyevka, and other areas. A total of 12 oil and gas fields discovered worth $ 4.38 billion in the prices of 1991 and $ 26.3 billion in the prices of 2008. For the discovery of these new oil and gas accumulations *I.I.Chebanenko, V.A.Krayushkin, V.P.Klochko, E.S. Dvoryanin, V.V.Krot,*

P.T.Pavlenko, M.I.Ponomarenko, and *G.D.Zabello,* were awarded the State Prize of Ukraine in the Field of Science and Technology in 1992 [*Chebanenko et al.,* 2002].

Today there are 50 commercial gas and oil fields known in the DDB's northern flank. Data obtained from drilling in many of these areas shows that the crystalline basement of northern flank consists of amphibolites, charnockites, diorites, gneisses, granites, granodiorites, granito-gneisses, migmatites, peridotites, and schists. 32 of commercial fields have oil and/or gas accumulations in sandstones of the Middle and Lower Carboniferous age. 16 other fields contain reservoirs in the same sandstones but separately from them - in amphibolites, granites and granodiorites of crystalline basement as well. Two fields contain oil pools in the crystalline basement only.

An exploration drilling in the DDB's northern flank has discovered five petroleum reservoirs in the Precambrian crystalline basement rock complex at the depths ranging from several meters to 336 meters below the top of crystalline basement. Gas- and oil-shows have been found in the Precambrian crystalline basement rocks as deep as 760 m below the top of crystalline basement. The seal rock for the reservoirs in the Carboniferous period sandstones are shallower shale formations. This is typical for petroleum pools in sedimentary beds. The cap rock for the reservoirs in the Precambrian crystalline basement is the impervious, non-fractured, essentially horizontal, layer-like zones of crystalline rock which alternate with the fractured, un-compacted, bed-like zones of granite, amphibolite and the other crystalline rocks mentioned above [*Krayushkin et al.,* 2001].

The exploration drilling in the DDB's northern flank is still in progress and continues to yield success in the 100x600 km petroliferous strip of the DDB's northern flank. Its proven petroleum reserves are already equal to $289 \cdot 10^6$ tons ($230 billion at 50 USD/bbl oil prices). The DDB's northern flank is even more attractive with its total perspective "in place" petroleum resources which amounts to about $13{,}000 \cdot 10^6$ tons (~80,000 bbls) of oil equivalent in an area of 48,000 sq km. The petroleum potential of the DDB's southern flank should not be neglected either with a total "in place" prognostic petroleum resources of about $6000 \cdot 10^6$ tons of oil equivalent in an area of 22,000 sq km. Here several promising leads with oil-shows can be found in the crystalline basement and its sedimentary cover [*Chebanenko et al.,* 1996].

Abyssal abiogenic petroleum has been discovered in China as well: the giant Xinjiang gas field contains about $400 \cdot 10^{12}$ m^3 of abiogenic natural gas [*Zhang,* 1990]. Chinese petroleum geologists estimated this quantity in volcanic island arcs, trans-arc zones of mud volcanism, trans-arc rift basins, trans-arc epicontinental basins, deep fault zones and continental rift basins.

Conclusions:

1. According to the traditional biotic petroleum origin hypothesis the DDB's northern flank was qualified as possessing no potential for petroleum production.
2. Based on the theory of the abyssal abiogenic origin of petroleum 50 commercial gas and oil deposits were discovered in this area. This is the best best evidence confirming the theory.

9. Deep and ultra-deep petroleum reservoirs

In this part of the chapter we discuss how far the distribution, location and reservoir conditions in the deep and ultra-deep petroleum deposits can be explained by the traditional biotic petroleum origin. The key points are as follows:

- deep and ultra-deep petroleum fields are below "the main zone of petroleum formation" determined by the traditional biotic petroleum origin hypothesis i.e the depth of 2-4 km and in exceptional cases – on the depth down to maximum 8 km.
- reservoir temperature of these fields is much higher than the optimal temperature range of the traditional biotic hypothesis of petroleum formation;
- the biotic hypothesis suggests that with growing depth and temperature hydrocarbons are destructed, reservoir rock porosity drops, thus petroleum reserves should be significantly reduced. A presence of more than 1000 petroleum deposits at the depth 5-10 km all over the world contradicts to these points –a seen below:

There are more than 1000 commercial petroleum fields producing oil and/or natural gas from sedimentary rocks at the depths of 4500-10,685 m. These fields were discovered in 50 sedimentary basins over the world.

Russia. A number of oil-and-gas fields have been discovered in the depth of 4000-4600 m in Russia. The cumulative production of these fields is equal to $421 \cdot 10^6$ t of oil, $45.5 \cdot 10^9$ m^3 of the associated oil gas and $641 \cdot 10^6$ m^3 of natural gas. Although these fields are not "ultradeep" reservoirs, but they are interesting from our points of view: they are associated with deep faults intersecting the whole sedimentary rock sequence. The "roots" of these deep faults extend underneath the basement part of lithosphere. The roots form vertical columns-("pipes") of high permeability/petroleum saturation and chains of oil and gas accumulations are connected to them. It was established that the traces of petroleum migration are entirely absent outside of anticline crests [*Istratov*, 2004].

Ukraine. 17 giant and supergiant gas fields were discovered in the Lower Carboniferous age sandstones of the Dnieper-Donets Basin at the depth range of 4500-6287 m. At these depths, the total proven reserves of natural gas is $142.6 \cdot 10^9$ cu m. The total recoverable reserves of condensate is $2.3 \cdot 10^6$ t [*Gozhik et al.*, 2006].

U.S.A. In the U.S.A. more than 7000 boreholes with TD deeper than 4575 m were drilled between 1963-1979. In the Mesozoic-Cenozoic rift system of the Gulf of Mexico a regional of the deep-to-ultra-deep (the Upper Cretaceous) sands have been observed. With a width of 32-48 km and a length of 520 km this trend extends along the Gulf of Mexico from New Orleans to the borderline between Louisiana and Texas. Many oil and gas fields were found in this formation over the area indicated at the depths of 4500-6100 m. Most of them have anomalously high reservoir temperature (Freeland field: 232 degrees C) what is much higher than the optimal temperature of petroleum formation from ancient organic materials. The total proven reserves of natural gas in the Tsucaloosa trend are equal to $170 \cdot 10^9$ m^3 but there is an opinion that only the central portion of it contains potential resources as much as

$850 \cdot 10^9$ m^3 of natural gas and $240 \cdot 10^6$ m^3 of condensate. [*King*, 1979; *Matheny*, 1979; *Pankonien*, 1979; *Sumpter*, 1979].

- In 2009 BP has discovered the Tiber crude oil deposit in the Mexican Gulf. The well, located in Keathley Canyon block 102, approximately 400 kilometres south east of Houston, is in 1,259 metres of water. The Tiber well was drilled to a total depth of 10,685 metres. The Tiber deposit holds between 4 billion and 6 billion barrels of oil equivalent, which includes natural gas (the third biggest find in the US) (http://www.bp.com/genericarticle.do?categoryId=2012968&contentId=7055818).

10. Supergiant oil and gas accumulations

One of the main problems of the traditional biotic petroleum origin hypothesis is the identification of biotic sources and material balance of the hydrocarbon generation for most of supergiant oil and gas fields.

Middle East. In the Middle East proven recoverable reserves are equal to $101 \cdot 10^9$ t of oil and $75.8 \cdot 10^{12}$ m^3 of gas respectively as of the end 2010 [*BP*, 2011]. Saudi Arabia's proven reserves are $36 \cdot 10^9$ t of oil and $8 \cdot 10^{12}$ m^3 of natural gas [*BP*, 2011]. Most of these reserves are located in ten supergiant gas and oilfields (Table 6) [*International*, 1976; *Alhajji*, 2001, *The List*, 2006]. These giant oil fields give oil production from the Jurassic-Cretaceous granular carbonates. All these crude oils have very similar composition referring to a common source. Such source is the Jurassic-Cretaceous thermally mature, thin-bedded organic rich carbonate sequence (3-5 mass %). Organic material is concentrated in dark, 0.5-3.0 mm thin beds alternating with the lightly colored, similarly thin beds poor in organics. Let's make a calculation of the oil might have been generated inside the basins of Saudi Arabia with an estimated "original-oil-in-place" of $127 \cdot 10^9$ m^3 [*BP*, 2011]. Areas within the sedimentary basins where the kerogen is mature (i.e. H/C ratio is 0.8-1.3), were mapped [*Ayres et al.*, 1982] and multiplied by the thickness of the source zones. This simple calculation gives a volume of petroleum source rocks as high as 5000 cubic km. If we accept that

- the volume of kerogen is equal to 10 % of the petroleum source rocks volume,
- the coefficient of transformation of kerogen into bitumen is equal to 15 %,
- that 10 % of this bitumen can migrate out of the petroleum source rocks,

we come to the conclusion, that only $7.5 \cdot 10^9$ m^3 of oil could migrate out of the petroleum source rocks. This is less than 6 % of Saudi Arabia's estimated "in place" oil reserves. Note, that if the kerogen transformation parameters are twice as high as taken here (i.e. 20%, 30% and again 20% respectively), the OOIP is still $60 \cdot 10^9 m^3$ i.e. half of the booked value.

Where did 94 % of Saudi Arabia's recoverable oil come from? This question is not a rhetorical one because any other source of beds of petroleum is absent in the subsurface of Saudi Arabia as well as of all countries mentioned above, according to *Ayres et al.* [1982], and *Baker et al.* [1984]. Bahrain, Iran, Iraq, Kuwait, Oman, Qatar, Saudi Arabia, Syria, United Arab Emirates and Yemen occur in the same sedimentary basin – the Arabian-Iranian Basin, where *Dunnington* [1958; 1967] established the genetic relationship, i.e. the single common source of all crude oils.

Canada. The West Canadian sedimentary basin attracts a great attention also. There is the unique oil/bitumen belt extended as the arc-like strip of 960 km length from Peace River through Athabasca (Alberta Province) to Lloydminster (Saskatchewan Province). This belt includes such supergiant petroleum fields as Athabasca (125 km width, 250 km length), Cold Lake (50 km, 125 km), Peace River (145 km, 180 km) and Wabaska (60 km, 125 km). Here the heavy (946.5-1.029 kg/m^3) and viscous (several hundred-several million cP) oil saturates the Lower

Deposit	*Estimated reserves*
Ghawar	$10.2 \cdot 10^9$-$11.3 \cdot 10^9$ t of oil $1.5 \cdot 10^{12}$ m^3 of gas
Safaniyah	$4.1 \cdot 10^9$ t of oil
Shaybah	$2.0 \cdot 10^9$ t of oil
Abqaiq	$1.6 \cdot 10^9$ t of oil
Berri	$1.6 \cdot 10^9$ t of oil
Manifa	$1.5 \cdot 10^9$ t of oil
Marjan	$1.4 \cdot 10^9$ t of oil
Qatif	$1.3 \cdot 10^9$ t of oil
Abu Safah	$0.9 \cdot 10^9$ t of oil
Dammam	$0.7 \cdot 10^9$ t of oil

Table 6. Supergiant Oil and Gas Deposits in Saudi Arabia

Cretaceous sands and sandstones. These fields contain the "in place" oil reserves equal to $92 \cdot 10^9$-$187 \cdot 10^9$ m^3 in Athabasca, $32 \cdot 10^9$-$75 \cdot 10^9$ m^3 in Cold Lake, $15 \cdot 10^9$-$19 \cdot 10^9$ m^3 in Peace River, $4.5 \cdot 10^9$-$50 \cdot 10^9$ m^3 in Wabaska, and $2 \cdot 10^9$-$5 \cdot 10^9$ m^3 of oil/bitumen in Lloydminster, totally $170 \cdot 10^9$-$388 \cdot 10^9$ m^3 [*Vigrass*, 1968; *Wennekers*, 1981; *Seifert et al.*, 1985; *Sincrude*, 1992; *Warters et al.*, 1995].

The conventional understanding is, that the oil of Athabasca, Cold Lake, Lloydminster, Peace River and Wabasca generated from dispersed organic matter buried in the argillaceous shales of the Lower Cretaceous Mannville Group only. It is underlain by the Pre-Cretaceous regional unconformity and its thickness varies from 100 to 300 m. Its total volume is about $190 \cdot 10^3$ cubic km with a 65 % shale content. Having the data of the total organic carbon concentration (TOC), the hydrocarbon index (HI), constant of transformation (K), and all other values from the accepted geochemical model of the oil generation from the buried organic matter dispersed in clays-argillites, it was concluded that the Mannville Group could only give $71.5 \cdot 10^9$ m^3 of oil. It is in several times less than the quantity of oil (see above) which was totally estimated before 1985 in Athabasca, Cold Lake, Lloydminster, Peace River and Wabasca oil sand deposits [*Moshier et al.*, 1985].

If we accept other estimations of the volume of oil/bitumen "in place" in Athabasca, Wabasca, Cold Lake and Peace River area (~ 122,800 sq km) conducted by Alberta Energy Utilities Board (AEUB) and the National Energy Board (NEB), Canada the gap between the booked and organically generated quantities is even wider (Table 12). AEUB estimated $270 \cdot 10^9$ m^3 of bitumen "in place", while NEB – $397 \cdot 10^9$ m^3 [*Canadian*, 1996].

In the above-mentioned area there additionally are $200 \cdot 10^9$-$215 \cdot 10^9$ m^3 of heavy (986-1030 kg/cu m) and viscous (10^6 cP under 16 degrees C) oil at the depth range of 75-400 m in the Upper Devonian carbonates (Grosmont Formation). They occur in the area of $70 \cdot 10^3$ km^2 beneath the Athabasca/Cold Lake/Lloydminster/Peace River/Wabasca oil sand deposits [*Wennekers*, 1981; *Seifert et al.*, 1985; *Hoffmann et al.*, 1986].

The total estimated reserves of bitumen "in place" in the above-mentioned area are between $370 \cdot 10^9$ and $603 \cdot 10^9$ m^3. If besides the Mannville clays and shales which could give $71.5 \cdot 10^9$ m^3 of oil only there is no any other petroleum source rock, where is a biotic source for the rest of 82-88% of oil in this area?

Venezuela. Something similar can be observed in the Bolivar Coastal oil field in Venezuela. According to *Bockmeulen et al.* [1983] the source rock of petroleum here is the La Luna limestone of the Cretaceous age. The estimated oil reserves are equal to $4.8 \cdot 10^9$ m^3 [*The List*, 2006] with an oil density of 820-1000 kg/m^3. The same kind of calculations that were done for Saudi Arabia above gives us the following result. One m^3 of the oil-generating rock contains $2.5 \cdot 10^{-2}$ m^3 of kerogen which can generate $2.5 \cdot 10^{-3}$ m3 bitumen giving $1.25 \cdot 10^{-4}$ m^3 of oil within the accepted geochemical model of biotic petroleum origin. Having this oil-generating potential and the $4.8 \cdot 10^9$ m^3 of estimated oil reserves of the Bolivar Coastal field as a starting point, the necessary volume of oil source rock would be equal to $3.84 \cdot 10^{13}$ m^3. This is consistent with the oil-generating basin area of 110 km across *if* the oil source rock is 1,000 m thick. The average thickness of La Luna limestone is measured with only 91 m [*Bockmeulen et al.*, 1983). The diameter of the oil-generating basin would be therefore equal to 370 km and area of this basin is equal approximately 50% of the territory of Venezuela what is geologically highly un-probable.

11. Gas hydrates: the greatest source of abiogenic hydrocarbons

Gas hydrates are clathrates. Looking as ice they consist of gas and water where the molecules of hydrate-forming gas (e.g., Ar, CH_4, C_2H_6, C_3H_8, *i*-C_4H_{10}, Cl, CO, CO_2, He, H_2S, N_2, etc.) are squeezed under the pressure of 25 MPa and more into the interstices of the water (ice) crystalline cage without any chemical bonding between molecules of water and gas. As a result thawing 1 m^3 of gas hydrate at the sea level produces 150-200 m^3 of gaseous methane and 0.87 m^3 of fresh water. Naturally, the formation of gas hydrates takes place under the great velocity of fluid movement and under the certain combination of pressure and temperature. For example, methane hydrate arises under conditions of -236 degrees C and $2 \cdot 10^{-5}$MPa; 57 degrees C and 1,146 MPa (*Klimenko*, 1989; *Makogon*, 1997; *Lowrie et al.*, 1999; *Makogon et al.*, 2005). There are also the data that the formation of gas hydrate from the

$CH_4/C_3H_8/CO_2/H_2O/H_2S$-mixture proceeds under such the high temperature increasing/such the high pressure decreasing that the gas hydrate of composition, above, arises and exists really in sea-bottom sediments where the depth of sea is only 50 m, e.g., in Caspian Sea [*Lowrie et al.*, 1999].

Visually the gas hydrates ("combustible ice") are the aggregate growths of transparent and semitransparent, white, gray or yellow crystals. They can partially or entirely saturate the natural porous media, adding the mechanical strength and acoustic hardness to sediments and sedimentary rocks. Boreholes and seismic surveys have established that methane hydrates occur in the polar regions of Asia, Europe, North America (fig. 10), where the "combustible ice" always is underlain with natural gas [*Trofimuk et al.*, 1975; *Panaev*, 1987; *Collett*, 1993; *Dillon et al.*, 1993; *Kvenvolden*, 1993; *Modiris et al.*, 2008].

The gas hydrates represent a huge unconventional resource base: it may amount from $1.5 \cdot 10^{16}$ m^3 [*Makogon et al.*, 2007] to $3 \cdot 10^{18}$ m^3 [*Trofimuk et al.*, 1975] of methane. Thus, natural gas reserves which could be obtained from "combustible ice" are enough for energy support for our civilization for the next several thousand years.

Top of the supergiant gas hydrate/free natural gas accumulations occur at a depth of 0.4-2.2 m below the sea bottom in the Recent sediments of the world ocean. The bottom of these accumulations is sub-parallel to the sea-bottom surface and intersects beds with anticlinal, synclinal, and tilted forms. This geometry, the geographical distribution of hydrates in the world ocean, their Recent to Pleistocene age and fresh water nature of "combustible ice" could not be explained by terms (source rocks, diagenesis and katagenesis/metagenesis of any buried, dispersed organic matter, lateral migration of natural gas) used in the traditional biotic petroleum origin hypothesis.

According to the theory of the abyssal abiogenic origin of petroleum all gas hydrate/free natural gas accumulations were formed due to the "one worldwide act" i.e. an upward vertical migration of abyssal abiogenic mantle fluid through all the faults, fractures, and pores of rocks and sea-bottom sediments. In that time, not more than 200 thousand years ago those faults, fractures, and pores were transformed by a supercritical geo-fluid (mixture of supercritical water and methane) into a conducting/accumulating/intercommunicating media. Acting as the natural "hydrofracturing" the abyssal geo-fluid has opened up the cavities of cleavage and interstices of bedding in the rocks and sediments as well. According to *Dillon et al.* [1993] the vertical migration of natural gas still takes place today on the Atlantic continental margin of the United States. Along many faults there the natural gas continues to migrate upwards through the "combustible ice" as through "sieve" that is distinctly seen as the torch-shape vertical strips in the blanking of seismographic records.

On April 10, 2012 Japan Oil, Gas and Metals National Corporation and ConocoPhillips have announced a successful test of technology dealing with safely extract of natural gas from methane hydrates. A mixture of CO_2 and nitrogen was injected into the formation on the North Slope of Alaska. The test demonstrated that this mixture could promote the production of natural gas. This was the first field trial of a methane hydrate production methodology whereby CO_2 was exchanged in situ with the methane molecules within a methane hydrate structure.

12. Conclusions

1. Geological data presented in this chapter do not respond to the main questions related to the hypothesis of biotic petroleum origin. Only the theory of the abiogenic deep origin of hydrocarbons gives convincing explanation for all above-mentioned data.
2. The experimental results discussed in the chapter confirm that the $CaCO_3$-FeO-H_2O system spontaneously generates the suite of hydrocarbons in characteristic of natural petroleum. Modern scientific considerations about genesis of hydrocarbons confirmed by the results of experiments and practical results of geological investigations provide the understanding that a part of the hydrocarbon compounds could be generated at the mantle conditions and migrated through the deep faults into the Earth's crust where they are formed oil and gas deposits in any kind of rocks and in any kind of their structural position.
3. The experimental results presented place the theory of the abiogenic deep origin of hydrocarbons in the mainstream of modern physics and chemistry and open a great practical application. The theory of the abiogenic deep origin of hydrocarbons confirms the presence of enormous, *inexhaustible* resources of hydrocarbons in our planet, allows us to develop a new approach to methods for petroleum exploration and to reexamine the structure, size and location of the world's hydrocarbons reserves (www.jogmec.go.jp).

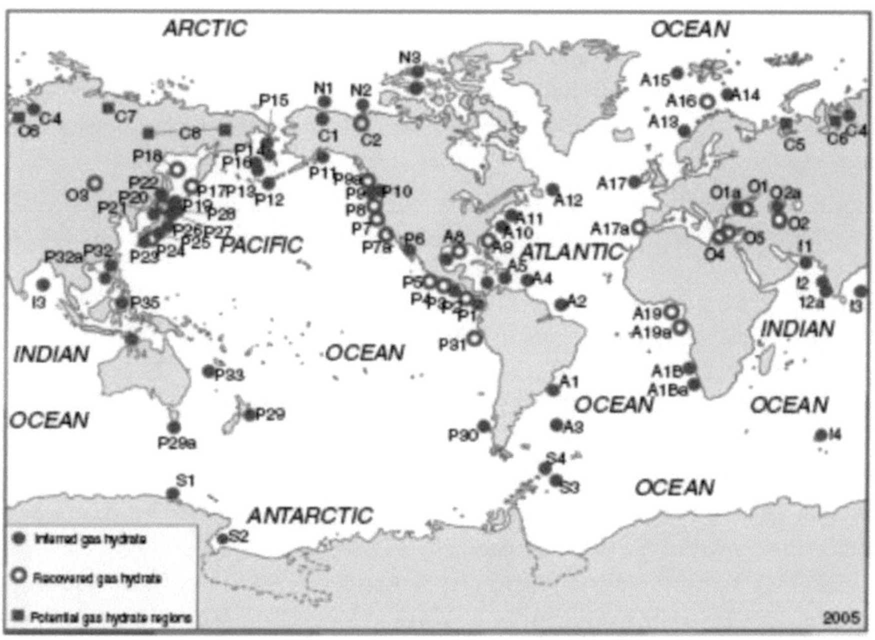

Figure 5. Inferred (63), recovered (23), and potential (5) hydrate locations in the world [*Kvenvolden and Rogers*, 2005].

Author details

Vladimir G. Kutcherov
Division of Heat and Power Technology, Royal Institute of Technology, Stockholm, Sweden
Department of Physics, Gubkin Russian State University of Oil and Gas, Moscow, Russia

13. References

Ayres, M. G., M. Bilal, R. W. Jones, L. W. Slentz, M. Tartir, and A. O. Wilson (1982), Hydrocarbon habitat in main producing areas, Saudi Arabia, *Amer. Assoc. Petrol. Geol. Bull.*, *66*, 1, 1-9.

Baker, E. G. (1962), Distribution of hydrocarbons in petroleum, *Amer. Assoc. Petrol. Geol. Bull.*, *46*, 1, 76-84.

Baker, C., and P. A. Dickey (1984), Hydrocarbon habitat in main producing area, Saudi Arabia, Discussion, *Ibid.*, *68*, 1, 108-109.

Blanc, G., J. Boulegue, and J. L. Charlou (1990), Profils d'hydrocarbures légers dans l'eau de mer, les saumures et les eaux intersticielles de la fosse Atlantis II (Mer Rouge), *Oceanol. Acta*, *13*, 187-197.

Bockmeulen, H., C. Barker, and P. A. Dickey (1983), Geology and geochemistry of crude oil, Bolivar Coastal fields, Venezuela, *Amer. Assoc. Petrol. Geol. Bull.*, *67*, 242-270.

BP Statistical Review of World Energy. June 2011,
http://www.bp.com/sectionbodycopy.do?categoryId=7500&contentId=7068481.

Brooks, J. M. (1979), Deep methane maxima in the Northwest Caribbean Sea: possible seepage along Jamaica Ridge, *Science*, *206*, 1069-1071.

Carson, R. W., D. G. Pearson, and D. E. James (2005), Physical, chemical, and chronological characteristics of continental mantle, *Rev. Geophys.*, *43*, RG1001.

Charlou, J. L., and J. P. Donval (1993), Hydrothermal methane venting between 12°N and 26°N along the Mid-Atlantic Ridge, *J. Geophys. Res.*, *98*, 9625-9642.

Charlou, J. L., J. P. Donval, Y. Fouquet, P. Jean-Baptist, and N. Holm (2002), Geochemistry of high H_2 and CH_4 vent fluids issuing from ultramafic rocks at the Rainbow hydrothermal field (36°14'N, MAR), *Chem. Geol.*, *184*, (1-2), 37-48.

Chebanenko, I. I., V. A. Krayushkin, V. P. Klochko, and V. P. Listkov (1996), Petroleum potential of non-traditional targets in Ukraine, *Geol. J.*, *3-4*, 7-11.

Chebanenko, I. I., P. F. Gozhik, V. A. Krayushkin, V. P. Klochko, B. J. Mayevsky, N. I. Yevdoshchuk, V. V. Gladun, T. E. Dovzhok, P. Ya. Maksimchuk, A. P. Tolkunov, O. G. Tsiokha, and M. G. Yegurnova (2005), *Petroleum in the Basement of Sedimentary Basins*, EKMO Publishing House, Kiev, Ukraine.

Chepurov A.I. *et al.*, (1999), Experimental study of intake of gases by diamonds during crystallization, *Journal of Crystal Growth*, *198/199*, 963-967.

Collet, T. S. (1993), Natural gas production from Arctic gas hydrates, in *The Future of Energy Gases*, pp. 299-311, US Geol. Survey Profes. Piper 1570, United States Government Printing Office, Washington (D.C.).

Craig, H., Y. Horibe, and K. A. Farley (1987), Hydrothermal vents in the Mariana Trough: Results of the first *Alvin* dives, *Eos Trans. AGU, 68*, 1531.

Curran, D. R., D. A. Shockley, L. Seaman, and M. Austin (1977), Mechanisms and models of cratering in earth media, in *Impact and Explosion Cratering*, pp. 1057-1087, Pergamon, Elmsford, NY.

Deines, P., J. W. Harris, P. M. Spear and J. J. Gurney (1989), Nitrogen and ^{13}C content of Finsh and Premier diamonds and their implication, *Ibid., 53*, 6, 1367-1378.

Dillon, W. P., M. W. Lee, and K. Fehlhalen (1993), Gas hydrates on the Atlantic continental margin of the United States – controls on concentration, in *Future of Energy Gases*, U.S. Geological Survey Professional Paper. 1570, pp. 313-330, edited by D.G.Howell, U.S. Govern. Print. Office, Washington (D.C.).

Donofrio, R. R. (1981), Impact craters: implications for basement hydrocarbons production, *Petrol. Geol. J., 3*, 3, 279-302.

Donofrio, R. R. (1998), North American impact structures hold giant field potential, *Oil and Gas J., 96*, 19, 69-80.

Dunnington, H. V. (1958), Generation, migration, accumulation, and dissipation of oil in Northern Iraq, in *Habitat of Oil*, pp. 1194-1251, Amer. Assoc. Petrol. Geol., Tulsa, OK.

Dunnington, H. V. (1967), Stratigraphic distribution of oil fields in the Iraq-Iran-Arabia Basin, *Inst. Petrol. J., 53*, 520, 129-161.

Evans, W. D., R. D. Norton, and R. S. Cooper (1964), Primary investigations of the oliferous dolerite of Dypvica, Arendal, S.Norway, in *Advances in Organic Geochemistry*, pp. 202-214, Pergamon, Oxford.

Giardini, A. A., Ch. E. Melton, and R. S. Mitchel (1982), The nature of the upper 400 km of the Earth and its potential as a source for nonbiogenic petroleum, *J.Petrol. Geol., 5*, 2, 130-137.

Gieskes, J. M., B. R. T. Simoneit, T. Brown, T. Shaw, Y. Ch. Wang, and A. Magenheim (1988), Hydrothermal fluids and petroleum in surface sediments of Guaymas Basin, Gulf of California: a case study, *Canad. Mineralogist, 26*, 3, 589-602.

Gozhik, P. F., V. A. Krayushkin, and V. P. Klochko (2006), Towards the search for oil at the depth of 8,000-12,500 m in the Dnieper-Donets Basin, *Geol. J., 4*, 47-54.

Green, D. H., W. O. Hibberson, I. Kovacs, and A. Rosenthal (2010), Water and its influence on the lithosphere–asthenosphere boundary. *Nature*, 467, 448-451, doi:10.1038/nature09369.

Grajales-Nishimura, J. M., E. Cedillo-Pardo, C. Rosales-Dominguez, D. J. Moran-Zenteno, W. Alvarez, Ph. Claeys, J. Ruiz-Morales, J. Garcia-Hernandes, P. Padilla-Avila, and A. Sanchez-Rios (2000), Chicxulub impact: The origin of reservoir and seal facies in the southeastern Mexico oil fields, *Geology, 28*, 4, 307-310.

Hazen R.M., R.J. Hemley and A.J. Mangum (2012), Carbon in Earth's interior: Storage, cycling, and life, *EOS, Transactions American Geophysical union, 93*, 2, 17, 2012, doi:10.1029/2012EO020001

Hoffmann, C. F., and O. P. Strausz (1986), Bitumen accumulation in Grosmont platform complex, Upper Devonian, Alberta, Canada, *Amer. Assoc. Petrol. Geol. Bull., 70*, 9, 1113-1128.

Horibe, Y. K., K. R. Kim, and H. Craig (1986), Hydrothermal methane plumes in the Mariana back-are spreading center, *Nature, 324*, 131-133.

India. Exploration and development (2006), *Oil and Gas J., 104*, 43, 46.

Irvine, T. N. (1989), A global convection framework: concepts of symmetry, stratification, and system in the Earth's dynamic structure?, *Econ. Geol., 84*, 8, 2059-2114.

Istratov, I. V. (2004), Chechnya sees potential for "nontraditional" deposits, *Oil and Gas J., 102*, 28, 38-41.

Kaminski, F. V., I. I. Kulakova, and A. I. Ogloblina (1985), About the polycyclic, aromatic hydrocarbons in carbonado and diamond, *Proceedings of AN of U.S.S.R, 283*, 985-988.

Kenney, J. F., V. G. Kutcherov, N. A. Bendeliani, and V. A. Alekseev (2002), The evolution of multicomponent systems at high pressures: YI. The thermodynamic stability of the hydrogen-carbon system: The genesis of hydrocarbons and the origin of petroleum, *Proceedings of the Nat. Acad. Sci, 99*, 17, 10976-10981.

King, R. E. (1979), Onshore activity surges in North America, *World Oil, 188*, 6, 43-57.

Klimenko, A. A., A. V. Botchkarev, and A. F. Nenakhov (1981), To a problem about the gas presence in the Paleozoic beds of the Scythian Plate, in *Questions to Drilling of Wells, Exploration and Exploation of Gas Fields in Northern Caucasus and Uzbekistan*, pp. 15-18, VNIIOUGP, Moscow.

Kolesnikov A., V.G. Kutcherov, A.F. Goncharov(2009), Methane-derived hydrocarbons produced under upper-mantle conditions, *Nature Geoscience, 2*, 566 – 570.

Koski, R. A., W. C. Shanks-III, W. A. Bohrson, and R. L. Oscarsen (1988), The composition of massive sulphide deposits from the sediment-covered floor of Escanaba Trough, Gorda Ridge: implication for depositional processes, *Canad. Mineralogist, 26*, 3, 655-673.

Krayushkin, V. A. (2000), The true origin, structure, patterns,and distribution of the world petroleum potential, *Georesources*, 3(4), 14-18.

Krayushkin, V. A., I. I. Tchebanenko, V. P. Klochko, Ye. S. Dvoryanin, and J. F. Kenney (2001), Drilling and development of the oil and gas fields in the Dnieper-Donets Basin, *Energia (Rivista Trimestrale sui prob-lemi dell'Energia), Anno XXII*, 3, 44-47.

Krayushkin, V. A., I. I. Tchebanenko, V. P. Klochko, B. B. Gladun, and O. G. Thiokha (2002), The basement petroleum presence in the Dnieper-Donets Basin, *Geology, Geophysics and Development of Oil and Gas Fields*, 1, 9-16.

Kulakova, I. I., A. I. Ogloblina, and V. I. Florovskaya (1982), Polycyclic, aromatic hydrocarbons in minerals-satillites of diamonds, and possible mechanism of their formation, *Proceedings of AN, U.S.S.R, 267*, 6, 1458-1460.

Kutcherov, V. G., N. A. Bendiliani, V. A. Alekseev, and J.F. Kenney (2002), Synthesis of hydrocarbons from minerals at pressure up to 5 GPa, *Proceedings of Rus. Ac. Sci., 387, 6, 789-792.*

Kutcherov V.G., A.Yu. Kolesnikov, T.I. Dyuzheva, L.F. Kulikova, N.N. Nikolaev, O.A. Sazanova, V.V. Braghkin (2010), Synthesis of Complex Hydrocarbon Systems at Temperatures and Pressures Corresponding to the Earth's Upper Mantle Conditions. *Doklady Akademii Nauk*, 433 (3), 361–364.

Kvenvolden, K. A., and B. R. T.Simoneit (1987), Petroleum from Northeast Pacific Ocean hydrothermal systems in Escanaba Trough and Guaymas Basin, *Amer. Assoc. Petrol. Geol. Bull., 71,* 5, 580.

Kvenvolden, K. A. (1993), Gas hydrates – geological perspective and global change, *Review of Geophysics, 31,* 2, 173-187.

Kvenvolden K.A., and B.W. Rogers (2005), Gaia's breath – global methane exhalations, *Marine and Petroleum Geology, 22,* 579-590.

Lonsdale, P. A. (1985), A transform continental margin rich in hydrocarbons, Gulf of California, *Amer. Assoc. Petrol. Geol. Bull, 69,* 7, 1160-1180.

Lowrie, A., and M. D. Max (1999), The extraordinary promise and challenge of gas hydrates, *World Oil, 220,* 9, 49-55.

Makeev, A. B., and V. Ivanukh (2004), Morphology of crystals, films, and selvages on a surface of the Timanian and Brazil diamonds, in *Problems of Mineralogy, Petrography and Metallogeny,* pp. 193-216, Perm. Univer. Press, Perm, Russia.

Makogon, Y. F. (1997). *Hydrates of Hydrocarbons,* PennWell, Tulsa.

Makogon, Y. F., S. A. Holdich, and T. Y. Makogon (2007), Natural gas-hydrates — A potential energy source

for the 21st Century, *Journal of Petroleum Science and Engineering, 56,* 14–31.

Masaitis, V. L., A. N. Danilin, and M. S. Mashchak (1980), *Geology of Astroblemes,* Nedra, Leningrad.

Masters, J. (1979), Deep Basin gas trap, West Canada, *Amer. Assoc. Petrol. Geol. Bull., 63,* 2, 152-181.

Matheny, Sh. L. (1979), Deeper production trend continuing, *Oil and Gas J., 77,* 15, 105-107.

Mao, Z., M. Armentrout, E. Rainey, C.E. Manning, P.K. Dera, V.B. Prakapenka, and A. Kavner (2011), Dolomite III: A new candidate lower mantle carbonate, *Geophys. Res. Lett.* 38, L22303, doi: 10.1029/2011GL049519

Melton, Ch. E., and A. A. Giardini (1974), The composition and significance of gas released from natural diamonds from Africa and Brazil, *Amer. Mineralogist, 59,* 7-8, 775-782.

Ditto (1975), Experimental results and theoretical interpretation of gaseous inclusions found in Arkansas natural diamonds, *Geochim. Cosmochim. Acta, 60,* 56, 413-417.

Moshier, S. O., and D. W. Wapples (1985), Quantitantive evaluation of Lower Cretaceous Manville group as source rock for Alberta's oil sands, *Amer. Assoc. Petrol, Geol. Bull., 69,* 2, 161-172.

Panaev, V. A. (1987), Gas hydrates in World Ocean, *Bull. Moscow Soc. Naturalists, Dep. Geol., 62,* 3, 66-72.

Pankonien, L. J. (1979), Operators scramble to tap deep gas in South Louisiana, *World Oil, 189,* 4, 55-62.

Peter, J. M., and S. D. Scott (1988), Mineralogy, composition, and fluid-inclusion microthermometry of seafloor hydrothermal deposits in the Southern Trough of Guaymas Basin, Gulf of California, *Canad. Mineralogist, 26,* 3, 567-587.

Proskurowski, G. *et al.* (2008), Abiogenic Hydrocarbon Production at Lost City Hydrothermal Field, *Science, 319,* 5863, 604-607

Ramboz, C., E. Oudin, and Y. Thisse (1988), Geyser-type discharge in Atlantis-II Deep, Red Sea: evidence of boiling from fluid inclusions in epigenetic anhydrite, *Canad. Mineralogist, 26*, 3, 765-786.

Rona, P. A. (1988), Hydrothermal mineralization at oceanic ridges, *Canad. Mineralogist, 26*, 3, 431-465.

Scott, H. P., R. J. Hemley, H. Mao, D. R. Herschbach, L. E. Fried, W. M. Howard, and S. Bastea, (2004), Generation of methane in the Earth's mantle: In situ high pressure-temperature measurement of carbonate reduction, *Proceedings Nat. Ac. Sci.,* 101, 39, 14023-14026.

Seifert, S. R., and T. R. Lennox (1985), Developments in tar sands in 1984, *Amer. Assoc. Petrol. Geol. Bull., 69*, 10, 1890-1897.

Simoneit, B. R. T. (1988) Petroleum generation in submarine hydrothermal systems: an update, *Canad. Mineralogist, 26*, 3, 827-840.

Simoneit, B. R. T., and P. F. Lonsdale (1982), Hydrothermal petroleum in mineralized mounds at the seabed of Guaymas Basin, *Nature, 295*, 5846, 118-202.

Sumtper, R. (1979), Strikes sparking action along Tuscaloosa trend, *Oil and Gas J., 77*, 43, 19-23.

Thompson, G., S. E. Humphris, and B. Shroeder (1988), Active vents and massive sulphides at 26°N (TAG) and 37°N (Snakepit) on the Mid-Atlantic Ridge, *Canad. Mineralogist, 26*, 3, 697-711.

The List: Taking Oil Fields Offline (2006), *Foreign Policy, 155*.

Trofimuk, A. A., N. B. Cherski, and V. P. Tsarev (1975), Resources of biogenic methane of the World Ocean, *Proc. AN Ud.S.S.R., 225*, 4, 936-943.

Vigrass, L. W. (1968), Geology of Canadian heavy oil sands, *Amer. Assoc. Petrol. Geol. Bull., 52*, 10, 1984-1999.

Warters, W. J., D. J. Cant, P. Tzeng, and P. J. Lee (1995), Western Canada Mannville gas potential seen as high, *Oil and Gas J., 93*, 45, 53-55.

Welhan, J. A., and H. Craig (1979), Methane and hydrogen in East Pacific Rise hydrothermal fluids, *Geophys. Res. Lett., 6*, 11, 829-831.

Wennekers, J. H. M. (1981), Tar sands, *Amer. Assoc. Petrol. Geol. Bull., 66*, 10, 2290-2293.

Zhang Kai (1990), The "in-place" resource evaluation and directions of exploration for the major natural gas accumulations of the non-biogenic origin in Xinjang, *Petrol. Exploration and Development, 17*, 1, 14-21.

Chapter 2

Thermogenic Methane with Secondary Alteration in Gases Released from Terrestrial Mud Volcanoes

Ryoichi Nakada and Yoshio Takahashi

Additional information is available at the end of the chapter

http://dx.doi.org/10.5772/48232

1. Introduction

Mud volcanoes are surface expressions of mud accompanied by water and gas originated from deep underground. They are found all over the world. The locations of mud volcanoes resemble magmatic volcanoes, that is, they are concentrated in areas of compressional tectonic settings such as accretionary complexes and subduction zones (Dimitrov, 2002, 2003; Kholodov, 2002; Kopf, 2002). Recent developments in seismic exploration and seafloor imaging have led to the discovery of mud volcanoes not only onshore, but also offshore. The fact that mud volcanoes are found along the compressional area suggests that eruptions are related to the occurrence of volcanic and earthquake activity. Mud extrusion is a phenomenon wherein fluid-rich, fine-grained sediments accompanying the gases ascend within a lithologic succession through conduits from pressurized reservoirs because of their buoyancy. The factors controlling the occurrence of mud volcanoes are considered to be (i) recent tectonic activity, particularly in a compressional regime; (ii) rapid loading of rocks due to fast sedimentation, accretion, or overthrusting; (iii) active hydrocarbon generation; and (iv) existence of thick, fine-grained, soft sediments deep in the sedimentary succession (Dimitrov, 2002). The main factor in mud volcano formation is considered to be a gravitative instability in low-density sediments below high-density rocks induced by fast sedimentation.

The major differences between mud volcano and normal (magmatic) volcano are as follows: (i) mud volcano only releases, as suggested by its name, mud associated with water, whereas magmatic volcano releases ash and high-temperature lava; and (ii) most of the gases released from the former are methane (CH_4), whereas the latter releases CO_2 and N_2, except for water vapor. With regard to difference (i), one may think that the mud volcano is not serious as a natural disaster. On the contrary, in Indonesia for example, more than 30,000 people lost their homes due to the eruption of mud (Mazzini et al., 2007). The eruption of an enormous amount of mud (170,000 m^3 per day at the maximum) with the temperature close to 100 °C buried the Sidoarjo village in Northeast Java (Mazzini et al., 2007). Thus, it is important to understand the eruption mechanism from the view of the disaster caused by the eruption. Difference (ii) is also important, considering that CH_4 has a larger global warming potential than CO_2 (IPCC, 2001). According to the IPCC report (2001), the global warming potential of CH_4 is 62 times higher than that of CO_2 in 20 years and 23 times higher in 100 years. The reported CH_4 concentration released from mud volcanoes all over the world shows that CH_4 dominates more than 90% for most mud volcanoes (Table 1). Furthermore, even if mud volcanoes are in the quiescent period, they constantly release gases into the atmosphere. Considering that magmatic volcanoes are not active in the quiescent period, the consecutive release of CH_4 from mud volcanoes is potentially an important problem. Therefore, understanding the source, abundance, and cause of CH_4 release from mud volcanoes is necessary to evaluate the global warming and potential resource as energy.

Both the concentration and CH_4 flux from mud volcanoes to the atmosphere are important. Thus far, several estimates for global emission have been done, including 10.3 Tg y^{-1} to 12.6 Tg y^{-1} (Dimitrov, 2002), 5 Tg y^{-1} to 10 Tg y^{-1} (Etiope and Klusman, 2002), 5 Tg y^{-1} (Dimitrov, 2003), and 6 Tg y^{-1} to 9 Tg y^{-1} (Etiope and Milkov, 2004). The estimates include several assumptions that can have large uncertainty in their flux estimation because it is almost impossible to determine the quantity of CH_4 released from each mud volcano on Earth. More recently, it has also been reported that gases from mud volcanoes not only originate from visible bubbling in the crater of mud volcanoes but also from soils around mud volcanoes. For example, Etiope et al. (2011) performed flux measurements from soils around mud volcanoes in Japan and showed that total output from soils is comparable with that from vents in the mud volcanoes. Their calculation suggests that global CH_4 flux from mud volcanoes is between 10 and 20 Tg y^{-1} (Etiope et al., 2011). These estimates mean that mud volcanoes represent an important natural source of atmospheric CH_4 considered in global greenhouse gas emission inventories.

Understanding the origin of CH_4, namely, microbial origin from acetate fermentation, microbial from carbonate reduction, thermogenic, and inorganic, provides information on the process and environment responsible for its generation. The interpretation of the origins of gas is generally based on its stable carbon and hydrogen isotopes ($\delta^{13}C$ and δD, respectively), and on the chemical composition of its gaseous alkanes (C_1–C_4; methane, ethane, propane, and butane). In particular, identifying the gas source is accomplished by plotting the stable carbon isotope ratio of C_1 ($\delta^{13}C_1$) versus the light gas composition (Bernard et al., 1978), and the $\delta^{13}C_1$ versus δD_1 (Schoell, 1983). Post-genetic alterations that can affect isotopic and molecular composition of gas should also be considered. The processes include (i) aerobic and anaerobic microbial oxidation of CH_4, (ii) abiogenic oxidation, (iii) isotopic fractionation by diffusion, (iv) molecular fractionation by advection, (v) gas mixing, and (vi) anaerobic biodegradation of petroleum and secondary methanogenesis. In this respect, both the chemical and isotopic compositions of hydrocarbons and of CO_2 can be useful. In this chapter, we attempt to improve our understanding of the origin of gases released from terrestrial mud volcanoes and seepages by summarizing published data. Further knowledge will allow researchers to use seepage gases as a tracer for hydrocarbon reservoirs and as an indicator of geodynamic processes, hazards, and importance in global changes.

2. Database

The database used in this chapter includes all terrestrial mud volcanoes and other seeps for which all the following parameters are reported: CH_4 stable isotopes ($\delta^{13}C_1$ and δD_1), compositional ratio of hydrocarbons [$C_1/(C_2 + C_3)$], and concentration and stable carbon isotope ratio of CO_2. The data which satisfy these restrictions are listed in Table 1. From more than 200 data, only 27 data from five countries consisted of all five parameters: 14 mud volcanoes from Azerbaijan, 7 from China, 1 from Georgia, 2 from Japan, and 3 from Turkmenistan; all other data lacked at least one parameter (Valyaev et al., 1985; Etiope et al., 2011; Nakada et al., 2011). Numerous studies have reported on at least one of the parameters above and/or the data of gases collected from the same mud volcanoes in different periods. However, the discussion should be performed using all the parameters above reported in one study, because (i) gases released from mud volcanoes have a complicated history, including secondary alterations, and (ii) compositions and stable isotope ratio can be fluctuated with time even in the same vent. Meanwhile, data from peats, recent sediments in freshwater environments, anthropogenically induced seeps from coal mines, coal-bed CH_4 production, and submarine mud volcanoes are not considered.

3. Results and discussion

3.1. The "Bernard" and "Schoell" diagrams

All the data listed in Table 1 are plotted in the "Bernard" and "Schoell" diagrams, namely, $\delta^{13}C_1$ versus $C_1/(C_2 + C_3)$ (Bernard et al., 1978; Faber and Stahl, 1984), and $\delta^{13}C_1$ versus δD_1 (Schoell, 1983). The former plot, which is widely used for the discrimination of thermogenic and microbial C_1, was originally developed by Bernard et al. (1978) through their analysis of hydrocarbons from Texas shelf and slop sediments. In 1984, Faber and Stahl collected sediment samples from the North Sea and modified the Bernard plot by adding the maturation trends of type II and type III kerogen. Figure 1 shows that all the gases released from mud volcanoes in China and Japan fall within or close to the thermogenic field. One of three data in Turkmenistan also falls within the thermogenic field, while two data from Turkmenistan and Georgia are in the intermediate region of the thermogenic and microbial fields. The rest of the data, all from Azerbaijan and one-third from Turkmenistan, fall in the region A, an ambiguous sector above the thermogenic field and right to the microbial. Gases from mud volcanoes do not appear to originate from microbial activities. However, gases from mud volcanoes in Azerbaijan, Italy, Papua New Guinea, and Russia (Taman Peninsula) fall in the microbial area (Valyaev et al., 1985; Baylis et al., 1997; Etiope et al., 2007). The data listed in Table 1 are selected ones that show all five parameters described in the previous section. Hence, the data lacking in other parameters, such as δD_1 or $\delta^{13}C_{CO2}$, are not considered in the present work. Then, it should be noted that not all the gases released from mud volcanoes are of thermogenic origin.

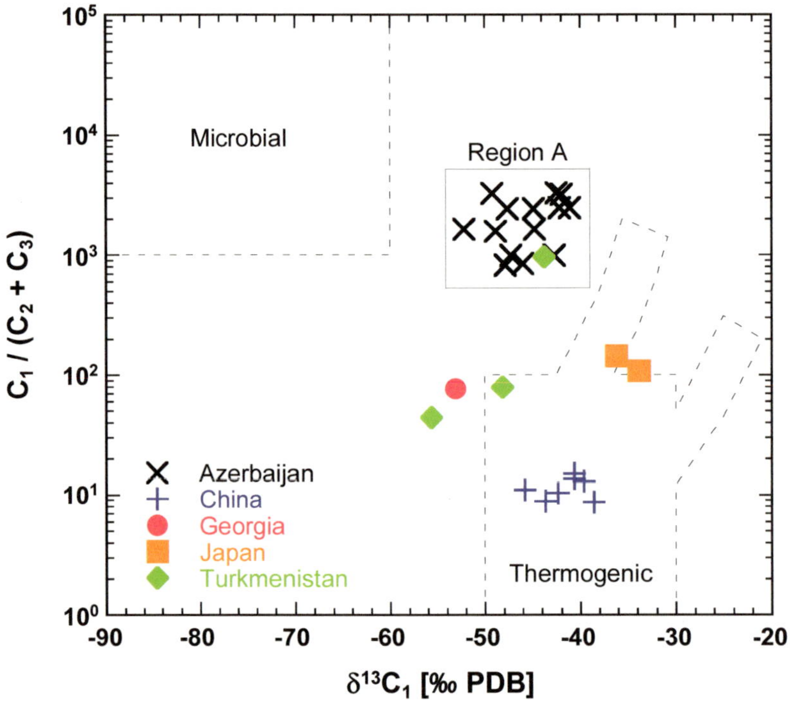

Figure 1. Carbon isotope ratio of CH_4 vs. hydrocarbon molecular composition diagram (Bernard plot; Bernard et al., 1978).

The "Schoell" plot, developed by Schoell (1983) through a summary of the genetic characterization of natural gases from several basins and areas including Gulf of Mexico, German Molasse, and Vienna (references therein), also shows that the data summarized here do not fall in the microbial field (Fig. 2). Likewise, no data are plotted in the dry thermogenic field (T_D). Most of the data fall in the thermogenic field associated with oil or the mixed field. The gases collected in Japan were plotted in the thermogenic field with condensate. Similar to the discussion for the Bernard diagram, the data selected in this work do not cover all the reported data on gases released from mud volcanoes. Actually, gases released from mud volcanoes in Papua New Guinea and Italy fall in the microbial field (Baylis et al., 1997; Etiope et al., 2007). However, until now, any combination of $\delta^{13}C_1$ and δD_1 is not reported for the gas samples that fall on the dry thermogenic area released from mud volcano, though gases from water seeps and dry seeps sometimes fall on the dry thermogenic field (Etiope et al., 2006; 2007, Greber et al., 1997).

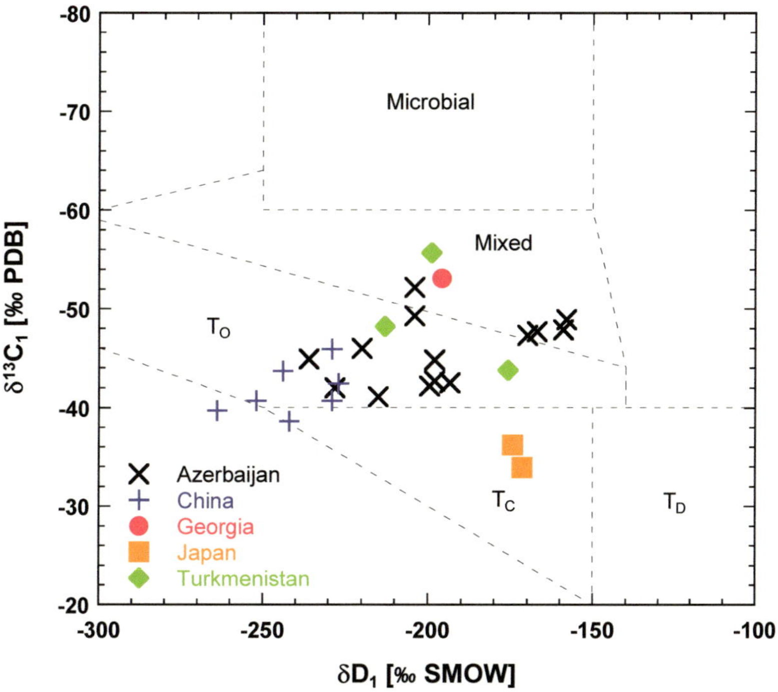

Figure 2. Carbon and hydrogen isotope diagram of CH_4 (Schoell plot; Schoell, 1983); To: thermogenic with oil; Tc: thermogenic with condensate; T_D: dry thermogenic.

The above figures suggest that thermogenic hydrocarbons are the main component of gases released from mud volcanoes. Considering that about half the gases from Azerbaijan fall in the thermogenic field in the Schoell plot, the reason gases plotted in region A in the Bernard diagram may be due to the fact that mixing between thermogenic and microbial or compositional ratio of hydrocarbons changed during post-genetic alteration. Considering $\delta^{13}C_1$ values alone, gases from Azerbaijan can be regarded as thermogenic, whereas Bernard ratios, namely, $C_1/(C_2 + C_3)$, are in the range of microbial origin. Therefore, it can be natural to consider that the data indicate mixing of the gases with two origins. When combining the origins, however, both Bernard ratios and $\delta^{13}C_1$ values are high enough to assume mixing between thermogenic and microbial, because the mixing trend generally tracks high $\delta^{13}C_1$ value with low Bernard ratio to low $\delta^{13}C_1$ value with high Bernard ratio and vise versa. This empirical rule suggests that gases from Georgia and one-third from Turkmenistan are regarded as tracking the mixing trend. On the other hand, gases from Azerbaijan and another one-third from Turkmenistan are not tracking the mixing trend, suggesting that the gases plotted in region A are not due to the mixing of thermogenic and microbial components. The data fall in region A in the Bernard diagram, therefore indicating post-genetic alteration such as (i) aerobic and anaerobic microbial oxidation of CH_4, (ii) abiogenic oxidation, and (iii) anaerobic biodegradation of petroleum and secondary methanogenesis. In this respect, discussion using only isotope and compositional ratios of hydrocarbon is not sufficient; CO_2 data provide useful information.

3.2. Large variation in $\delta^{13}C_{CO2}$ from mud volcanoes

In contrast to the $\delta^{13}C_1$ values, $\delta^{13}C_{CO2}$ from mud volcanoes show a large variation, from -36.9‰ to +29.8‰ (Table 1). Furthermore, the $\delta^{13}C_{CO2}$ values of the gases from Azerbaijan and China, which were plotted on a narrow range in the Bernard diagram, varied to a large degree. The $\delta^{13}C_1$ values of Azerbaijan show a variation of 11.1‰ (from -52.2‰ to -41.1‰), whereas the variation of $\delta^{13}C_{CO2}$ values is 51‰ (from -36.9‰ to +14.1‰). Similar to Azerbaijan, gases from seven mud volcanoes in China show a small variation in $\delta^{13}C_1$ (7.3‰, from -45.9‰ to -38.6‰) and a large variation in $\delta^{13}C_{CO2}$ (41.3‰, from -11.5‰ to +29.8‰). Besides the data selected there, similar characteristics also present in other reported data of gas released from mud volcanoes. Seven mud volcanoes in Georgia have a variation of 14.4‰ in $\delta^{13}C_1$ values and a 28.7‰ variation of $\delta^{13}C_{CO2}$ (Valyaev et al., 1985). Six vents of a mud volcano in Italy have a range of $\delta^{13}C_1$ in 4.3‰ with 29.1‰ variation of $\delta^{13}C_{CO2}$ values (Favara et al., 2001). Thirteen mud volcanoes with 20 reported isotopic ratios in Russia (Taman Peninsula) show a variation of 31.3‰ in $\delta^{13}C_1$ and a variation of 41.9‰ in $\delta^{13}C_{CO2}$ (Valyaev et al., 1985; Lavrushin et al., 1996). Twelve mud volcanoes with 15 vents in Trinidad display a variation of 21.6‰ in $\delta^{13}C_1$ and a 32.4‰ variation in $\delta^{13}C_{CO2}$ (Deville et al., 2003). Turkmenistan, with six available data of mud volcanoes, shows a variation of 12.5‰ in $\delta^{13}C_1$ with 32.1‰ variation in $\delta^{13}C_{CO2}$ (Valyaev et al., 1985). Eleven reported data from Ukraine show a 16.8‰ variation in $\delta^{13}C_1$ and a 40.9‰ variation in $\delta^{13}C_{CO2}$ (Valyaev et al., 1985). Seven mud volcanoes in Taiwan, in contrast, display a variation of 27.3‰ in $\delta^{13}C_1$ with 17.3‰ in $\delta^{13}C_{CO2}$ (Etiope et al., 2009). These facts clearly indicate that terrestrial mud volcanoes show a large variation in the isotopic ratio of CO_2, though most of their $\delta^{13}C_1$ values are within a thermogenic range.

In general, the $\delta^{13}C_{CO2}$ value ranges from -25‰ to -5‰ for natural thermogenic and/or kerogen decarboxylation (Jenden et al., 1993; Kotarba, 2001; Hosgormez et al., 2008). In addition, Jenden et al. (1993) suggested that the upper limit of $\delta^{13}C_{CO2}$ value due to the alteration of marine carbonates is +5‰. Therefore, CO_2 released from mud volcanoes with $\delta^{13}C_{CO2}$ value above that threshold can be called ^{13}C-enriched CO_2. Surprisingly, 14 mud volcanoes in the 28 listed in Table 1 release ^{13}C-enriched CO_2. Considering the 134 mud volcanoes described in the previous paragraph (data not shown), 66 (49%) of them show the ^{13}C-enriched value.

Before assessing the relationship between ^{13}C-enriched CO_2 and composition and isotopes of carbon in hydrocarbon gas, it is necessary to note that a large variability of the $\delta^{13}C_{CO2}$ value can be found within a mud volcano, both in space (gas samples from different vents) and in time (same vents analyzed in different time). For example, according to Nakada et al. (2011), four mud volcanoes (sites 1–4) are located very close to one another. The chemical compositions of mud and water, as well as relative abundances and stable isotopes of various hydrocarbons, are very similar. In particular, sites 2 to 4 are located within a 500 m distance, suggesting that their reservoir can be the same. Among these sites, however, only a gas released from site 2 has ^{13}C-enriched CO_2 (+16.2‰), whereas CO_2 from sites 1 and 4 are within a range of kerogen decarboxylation, -8.0‰ and -11.5‰, respectively (gases from site

Name	$\delta^{13}C_1$	δD_1	$\delta^{13}C_{CO2}$	CH_4 (%)	CO_2 (%)	$C_1/(C_2+C_3)$	Reference
Azerbaijan							
Airantekyan	–44.9	–236	+13.9	96.9	3.04	2423	Valyaev et al. (1985)
Akhtarma Pashaly	–47.9	–159	–7.2	99.1	0.68	825	Valyaev et al. (1985)
Chukhuroglybozy	–41.1	–215	+1.7	99.5	0.42	2488	Valyaev et al. (1985)
Dashgil	–42.2	–200	–6.4	99.0	0.93	2476	Valyaev et al. (1985)
Galmas	–47.7	–167	+13.7	97.4	2.06	2435	Valyaev et al. (1985)
Goturlyg	–42.7	–198	–15.4	98.9	0.99	989	Valyaev et al. (1985)
Gyrlykh	–49.3	–204	+0.6	98.4	1.54	3280	Valyaev et al. (1985)
Inchabel	–48.9	–158	–30.9	94.4	5.53	1573	Valyaev et al. (1985)
Kichik Kharami	–52.2	–204	+0.4	98.7	1.20	1646	Valyaev et al. (1985)
Shikhikaya	–47.3	–170	–36.9	98.9	0.99	989	Valyaev et al. (1985)
Shikhzagirli (Ilanly)	–42.5	–194	+0.1	99.1	0.81	3302	Valyaev et al. (1985)
Shokikhan	–42.0	–228	+13.8	96.7	3.30	3222	Valyaev et al. (1985)
Zayachya Gora (a)	–44.8	–198	+10.5	99.0	0.81	1647	Valyaev et al. (1985)
Zaakhtarma	–46.0	–220	+14.1	93.8	6.14	852	Valyaev et al. (1985)
China							
site 1	–45.9	–229	–8.0	91.6	0.20	11	Nakada et al. (2011)
site 2	–43.7	–244	+16.2	89.0	0.50	9	Nakada et al. (2011)
site 4	–42.4	–227	–11.5	89.6	0.10	10	Nakada et al. (2011)
site 6	–40.7	–252	+21.1	81.5	0.14	15	Nakada et al. (2011)
site 7	–39.7	–264	+24.6	80.4	0.17	13	Nakada et al. (2011)
site 8	–40.7	–229	+29.8	92.6	0.45	14	Nakada et al. (2011)
site 9	–38.6	–242	–4.8	75.8	0.31	9	Nakada et al. (2011)
Georgia							
Tyulkitapa	–53.1	–196	+5.9	89.0	10.86	77	Valyaev et al. (1985)
Japan							
Kamou	–33.9	–172	+10.9	95.4	2.91	108	Etiope et al. (2011)
Murono vent2	–36.2	–175	+28.3	93.7	5.62	144	Etiope et al. (2011)
Turkmenistan							
Keimir	–48.2	–213	–25.2	95.3	0.89	79	Valyaev et al. (1985)
Kipyashii Bugor	–43.8	–176	+6.9	96.8	2.79	968	Valyaev et al. (1985)
Ak-Patlauk	–55.7	–199	–16.2	94.2	3.67	44	Valyaev et al. (1985)

Table 1. Selected data of composition and stable isotope ratio of CH_4 and CO_2.

3 were not collected). Another example can be given by mud volcanoes in Japan. Kato et al. (2009) reported that the $\delta^{13}C_{CO2}$ value of gas released from the Murono mud volcano in August 2004 was +30.8‰. Mizobe (2007) showed that the value of the same mud volcano in May 2005 was +19.2‰ and in June 2006 was +21.2‰. Etiope et al. (2011) reported the value of the same mud volcano was +28.32‰ in May 2010. In contrast, the variation observed in $\delta^{13}C_1$ value of the Murono mud volcano reported in these papers was -33.1‰ to -36.2‰. These observations mean that the different vents of a mud volcano can be related to different circulation systems and/or post-genetic processes, and possibly different source pools or reservoirs. Some of the large Azerbaijan mud volcanoes show oil-saturated structures in some vents while others do not. This finding means that mud volcano systems may not be uniform, but can be structured in different systems and isolated blocks.

However, the variation of $\delta^{13}C_{CO2}$ with time for the same vent suggests that CO_2 carbon isotopes are intrinsically unstable and can be affected by multiple gas–water–rock interactions. According to the estimation by Pallasser (2000), however, the dissolution effect is limited in the carbon isotope enrichments of up to 5‰, suggesting that the main enrichment is due to biochemical fractionation related to secondary methanogenesis.

Figure 3 shows that ^{13}C-enriched CO_2 has no relation with CO_2 concentrations. Hypothetical end-members were assumed in the figure at 30% CO_2 with a carbon isotope ratio of 10‰ and 25% CO_2 with $\delta^{13}C_{CO2}$ of 30‰ for fermentation of hydrocarbon oxidation products, and 0% CO_2 with carbon isotope ratio of -20‰ and 0.5% CO_2 with $\delta^{13}C_{CO2}$ of 0‰ for thermogenic (Jeffrey et al., 1991). The observed data are distributed following the mixing trend between CO_2-rich gas produced by fermentation and CO_2-poor thermogenic gas. The two trend lines appear compatible with a mixing model and, therefore, with the presence of a residual CO_2 related to secondary methanogenesis and anaerobic biodegradation.

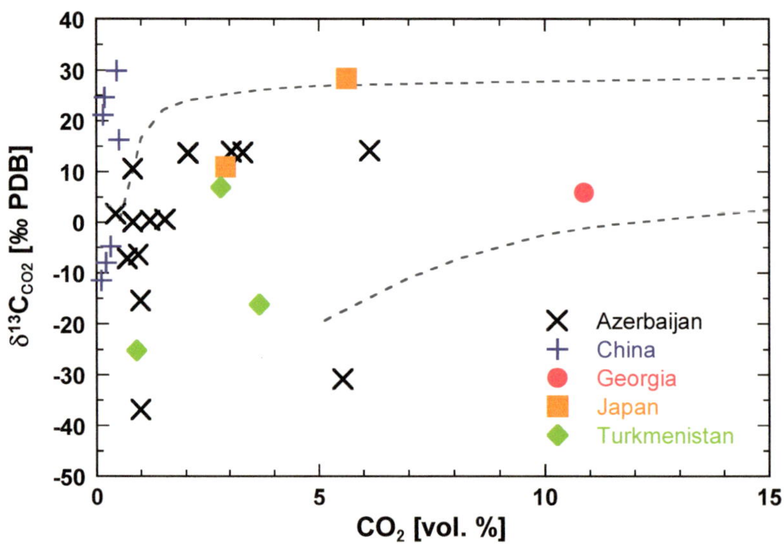

Figure 3. Relationship between $\delta^{13}C_{CO2}$ and CO_2 concentration. The 2 lines refer to a mixing trend similar to the model by Jeffrey et al. (1991).

3.3. CH_4 versus CO_2

The relationship between $\delta^{13}C$ of CH_4 and CO_2 is shown in Fig. 4. The ^{13}C-enriched CO_2 seems to occur preferentially in thermogenic CH_4, where $\delta^{13}C_1$ values are within a range of -50‰ to -30‰. In other words, this relationship seems to have a correlation; the gases showing a low $\delta^{13}C_1$ value have a low $\delta^{13}C_{CO2}$ value while a high $\delta^{13}C_1$ value corresponds to a high $\delta^{13}C_{CO2}$ value. This observation can imply that the light $\delta^{13}C_1$ in the few mud volcanoes (although data are not shown here, gases from Azerbaijan, Italy, Papua New Guinea, and Taman Peninsula fall in the microbial area in the Bernard diagram) with microbial gas is not due to secondary methanogenesis, but simply to primary methanogenesis. On the other hand, thermogenic gas with ^{13}C-enriched CO_2 maintains its high $\delta^{13}C_1$ value, indicating that $\delta^{13}C_1$ value is not perturbed by the secondary microbial gas. The $\delta^{13}C_1$ value does not vary to a large degree by post-genetic alteration because the amount of secondary microbial CH_4 is small compared with that of the pre-existing thermogenic gas.

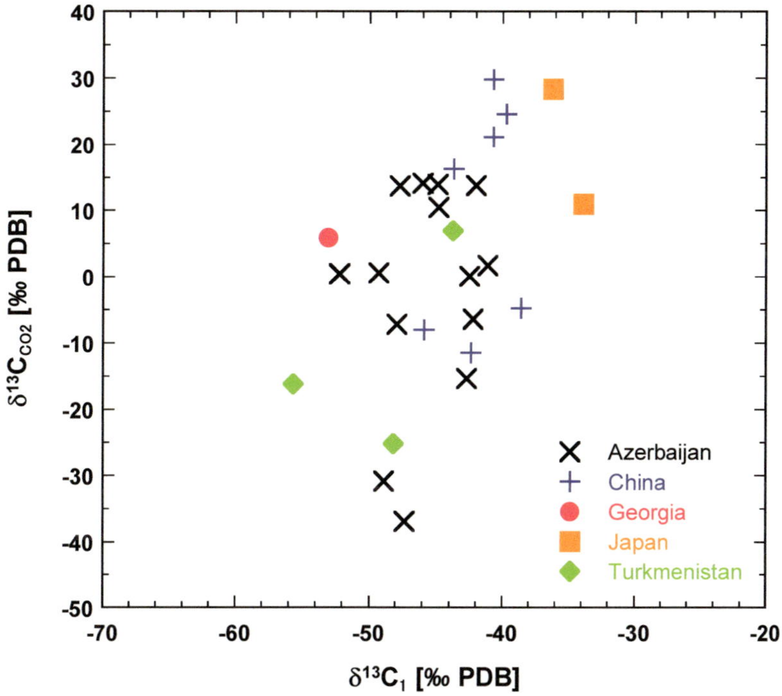

Figure 4. Relationship between carbon isotopes of CH_4 and CO_2.

However, the post-genetic alteration can lead to a significant change in their concentration and isotopic composition for CO_2. Oil biodegradation, one of the post-genetic alterations followed by CO_2 reduction, is described as follows:

$$CO_2 + 4H_2 \rightarrow CH_4 + 2H_2O. \quad (1)$$

This reaction is associated with a large kinetic isotope effect, meaning that the more the reaction proceeds with decreasing CO_2 concentration, the more ^{13}C is enriched in residual CO_2. Considering that CO_2 is a minor component of gases released from mud volcanoes, the concentration and isotopic composition of CO_2 can largely be affected by the reaction. The isotope effect of oil biodegradation results in the increase in $\delta^{13}C$ of residual CO_2, which can easily exceed +10‰ (Pallasser, 2000; Waseda and Iwano, 2008). Occurrence of oil biodegradation is suggested by the high C_2/C_3 and i-C_4/n-C_4 ratios (Pallasser, 2000; Waseda and Iwano, 2008) and/or by the presence of H_2 gas. For example, all these characteristics are identified in the gases released from mud volcanoes in China, which show a large variation in $\delta^{13}C_{CO2}$ as described above (Nakada et al., 2011). The increase of C_2/C_3 ratio due to oil biodegradation also leads to an increase of Bernard ratio, meaning that a gas sample plotted on the Bernard diagram will move in an upward direction. Thus, gases that fall in region A, which is geometrically above the thermogenic field in the Bernard diagram (Fig. 1), can be subject to the post-genetic alteration including biodegradation.

3.4. Depth of the reservoir and ^{13}C-enriched CO_2

The anaerobic biodegradation of oil and natural gas has been document to be mostly limited to shallow reservoirs, generally shallower than 2000 m with temperature below 60 °C to 80 °C (Pallasser, 2000; Feyzullayev and Movsumova, 2001). For example, the depth of petroleum reservoirs in the South Caspian Basin is shallower than 2000 m if the data are confined to the gas showing ^{13}C-enriched CO_2 (Fig. 5; after Etiope et al., 2009). This observation suggests that mud volcanoes showing anaerobic biodegradation signals will be linked with shallow reservoirs, while mud volcanoes without anaerobic biodegradation will more likely be produced by deeper reservoirs. Anaerobic biodegradation of petroleum and subsequent secondary methanogenesis, however, can also take place at shallower depths even above the deep reservoir along the seepage channels of the mud volcano system.

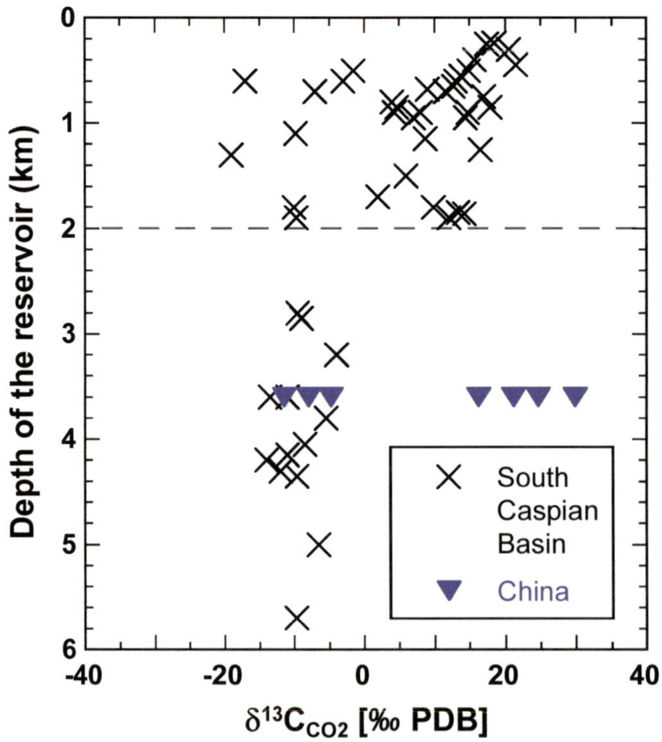

Figure 5. Carbon isotopic ratios of CO_2 vs. reservoir depth in the South Caspian Basin and China.

Recently, Nakada et al. (2011) showed that the reservoir depth of mud volcanoes in Xinjiang Province, China, is deeper than 3600 m, though some of the volcanoes release ^{13}C-enriched CO_2. The province hosts a large abundance of petroleum; therefore, many oil-testing wells are made in the province, leading to the knowledge of geothermal gradient and depth of oil reservoir. Nakada et al. (2011) calculated the equilibrium temperature of oxygen isotope fractionation between water and calcite in mud, indicating that the temperature where the water–rock interaction is occurring is 81 °C for mud volcanoes located close to the Dushanzi oil field. Assuming the mean geothermal gradient of the area is 18 ±1 °C/km (Nansheng et al., 2008) and that the surface temperature is 15 °C, the depth of the chamber with the temperature of 81 °C is calculated to be 3670 ±200 m. The calculated depth is slightly deeper than the oil reservoir at the Dushanzi field (3644 m–3656 m; Clayton et al., 1997). However, considering that the reservoir of saline fossil waters related to petroleum is generally deeper than that of oil and gas due to the difference in density, the calculation by Nakada et al. (2011) was surprisingly well consistent with the observation by Calyton et al. (1997). Then, Nakada et al. (2011) estimated the depth of gas reservoir at about 3600 m by considering that (i) the gases released from the mud volcanoes in the area were thermogenic gas associated with oil and (ii) the gas reservoir is generally located above the petroleum reservoir. The depth of 3600 m is greater than those previously reported for mud volcanoes releasing ^{13}C-enriched CO_2, such as those in South Caspian Basin. Thus, the secondary microbial effect that can occur at a relatively shallower depth must be considered separately from the initial thermogenic source in the field in China. Therefore, Nakada et al. (2011) clearly showed that the anaerobic biodegradation of petroleum can take place at a shallower depth. This result strengthens the model that considers a deep reservoir with thermogenic gas and secondary microbial activity occurring along the seepage system above the main deep reservoir.

4. Summary

Terrestrial mud volcanoes release a dominant abundance of thermogenic CH_4 related to the activities in relatively deep reservoirs, most of which are in petroleum seepage systems. Maturated petroleum associated with gas and water pressurizes the reservoir, causing gas and water to ascend preferentially through faults (Nakada et al., 2011). Some post-genetic secondary processes can alter the chemical and isotopic composition of the gases. Among these processes, some mixing, molecular fractionation, and particularly, secondary methanogenesis related to subsurface biodegradation of petroleum seem to be significant in changing the chemical and isotopic composition of gases released from mud volcanoes. Mud volcanoes show highly variable $\delta^{13}C_{CO2}$ values even within the same mud volcanoes, such that ^{13}C-enriched CO_2 can be found in some vents and not in others nearby, or not systematically changed in the same vent, meaning that ^{13}C-enriched CO_2 is, therefore, not an uncommon characteristic. The association of anaerobic biodegradation can depend on the type of microbial communities and physicochemical conditions of the reservoir.

Author details

Ryoichi Nakada* and Yoshio Takahashi
Department of Earth and Planetary Systems Science, Graduate School of Science, Hiroshima University, Higashi-Hiroshima, Hiroshima, Japan

Acknowledgement

R. N. is a Research Fellow of the Japan Society of the Promotion of Science. We thank Prof. Guodong Zheng (Chinese Academy of Sciences) for his suggestions throughout this study.

5. References

Baylis, S.A., Cawley, S.J., Clayton, C.J., Savell, M.A. (1997) The origin of unusual gas seeps from onshore Papua New Guinea. *Mar. Geol.*, 137, 109–120.

Bernard, B.B., Brooks, J.M., Sackett, W.M. (1978) Light hydrocarbons in recent Texas continental shelf and slope sediments. *J. Geophys. Res.*, 83, 4053–4061.

Capozzi, R., Picotti, V. (2002) Fluid migration and origin of a mud volcano in the Northern Apennines (Italy): the role of deeply rooted normal faults. *Terranova*, 14, 363–370.

Clayton, J. L., Yang, J., King, J. D., Lillis, P. G., and Warden, A. (1997) Geochemistry of oils from the Junggar Basin, Northwest China. *AAPG Bull.*, 81, 1926–1944.

Deville, E., Battani, A., Griboulard, R., Guerlais, S.H., Herbin, J.P., Houzay, J.P., Muller, C., Prinzhofer, A. (2003) Mud volcanism origin and processes. New insights from Trinidad and the Barbados Prism. In: Van Rensbergen, P., Hillis, R.R., Maltman, A.J., Morley, C. (Eds.), Surface Sediment Mobilization. Special Publication of the Geological Society (London), vol. 216, pp. 475–490.

Dimitrov, L. I. (2002) Mud volcanoes-the most important pathway for degassing deeply buried sediments. *Earth-Sci. Rev.*, 59, 49–76.

Dimitrov, L. I. (2003) Mud volcanoes-a significant source of atmospheric methane. *Geo-Mar. Lett.*, 23, 155–161.

Etiope, G. and Klusman, R. W. (2002) Geologic emissions of methane to the atmosphere. *Chemosphere*, 49, 777–789.

Etiope, G. and Milkov, A. V. (2004) A new estimate of global methane flux from onshore and shallow submarine mud volcanoes to the atmosphere. *Environ. Geol.*, 46, 997–1002.

Etiope, G., Papatheodorou, G., Christodoulou, D., Ferentinos, G., Sokos, E., Favali, P. (2006) Methane and hydrogen sulfide seepage in the NW Peloponnesus petroliferous basin (Greece): origin and geohazard. *AAPG Bull.*, 90, 701–713.

Etiope, G., Martinelli, G., Caracausi, A., Italiano. F. (2007) Methane seeps and mud volcanoes in Italy: gas origin, fractionation and emission to the atmosphere. *Geophys. Res. Lett.*, 34, L14303.

* Corresponding Author

Etiope, G., Feyzullayev, A., Milkov, A. V., Waseda, A., Mizobe, K. and Sun, C. H. (2009) Evidence of subsurface anaerobic biodegradation of hydrocarbons and potential secondary methanogenesis in terrestrial mud volcanoes. *Mar. Pet. Geol.*, 26, 1692–1703.

Etiope, G., Nakada, R., Tanaka, K., and Yoshida, N. (2011) Gas seepage from Tokamachi mud volcanoes, onshore Niigata Basin (Japan): Origin, post-genetic alterations and CH_4–CO_2 fluxes. *Appl. Geochem.*, 26, 348–359.

Faber, E., Stahl, W. (1984) Geochemical surface exploration for hydrocarbons in the North Sea. *AAPG Bull.*, 68, 363–386.

Favara, R., Gioia, C., Grassa, F., Inguaggiato, S., Proietto, F., Valenza, M. (2001) Studio geochimico delle manifestazioni fluide della riserva naturale integrale "Macalube di Aragona". *Nat. Sicil.*, 25, 137–154 (in Italian).

Feyzullayev, A. and Movsumova, U. (2001) About the origin of isotopically heavy CO_2 in gases of Azerbaijan mud volcanoes. *Azerb. Geol.*, 6, 96–105 (in Russian).

Greber, E., Leu, W., Bernoulli, D., Schumacher, M.E., Wyss, R. (1997) Hydrocarbon provinces in the Swiss southern Alps—a gas geochemistry and basin modeling study. *Mar. Pet. Geol.*, 14, 3–25.

Hosgormez, H., Etiope, G., and Yalçin, M. N. (2008) New evidence for a mixed inorganic and organic origin of the Olympic Chimaera fire (Turkey): a large onshore seepage of abiogenic gas. *Geofluids*, 8, 263–273.

Intergovernmental Panel on Climate Change (2001) Houghton, J.T., Ding, Y., Griss, D.J., Noguer, M., van der Linden, P.J., Dai, X., Maskell, K., Johnson, C.A. (Eds), Climate Change 2001: The scientific basis. Cambridge Univ PressCambridge, UK, pp. 881

Jeffrey, A.W.A., Alimi, H.M., Jenden, P.D. (1991) Geochemistry of the Los Angeles Basin oil and gas systems. In: Biddle, K.T. (Ed.), Active Margin Basins. AAPG Memoir, vol. 52, pp. 197–219.

Jenden, P. D., Hilton, D. R., Kaplan, I. R., Craig, H. (1993) Abiogenic hydrocarbons and mantle helium in oil and gas fields. In: Howell, D. (Ed.), The Future of Energy Gases, vol. 1570. USGS, pp. 31–35.

Kato, S., Waseda, A., Nishita, H., Iwano, H. (2009) Geochemistry of crude oils and gases from mud volcanoes and their vicinities in the Higashi-Kubiki area, Niigata Prefecture. *Journal of Geography*, 118, 455–471 (In Japanese with English abstract).

Kholodov, V. N. (2002) Mud volcanoes, their distribution regularities and genesis: communication 1. mud volcanic provinces and morphology of mud volcanoes. *Lithol. Miner. Resour.*, 37, 197–209.

Kopf, A. J. (2002) Significance of mud volcanism. *Rev. Geophys.*, 40, 1–52.

Kotarba, M. J. (2001) Composition and origin of coalbed gases in the Upper Silesian and Lublin basins, Poland. *Org. Geochem.*, 32, 163–180.

Lavrushin, V. Y., Polyak, B. G., Prasolov, R. M. and Kamenskii, I. L. (1996) Sources of material in mud volcano products (based on isotopic, hydrochemical, and geological data). *Lithol. Miner. Resour.*, 31, 557–578.

Mazzini, A., Svensen, H., Akhmanov, G. G., Aloisi, G., Planke, S., Malthe-Sørenssen, A., and Istadi, B. (2007) Triggering and dynamic evolution of the LUSI mud volcano, Indonesia. *Earth Planet. Sci. Lett.*, 261, 375–388.

Mizobe, K. (2007) Geochemical characteristics of natural gases from mud volcanoes in Tokamachi City, Niigata Prefecture. Master Thesis, Graduate School of Science and Engineering, Yamaguchi University, pp. 44.

Nakada, R., Takahashi, Y., Tsunogai, U., Zheng, G., Shimizu, H., and Hattori, K. H. (2011) A geochemical study on mud volcanoes in the Junggar Basin, China. *Appl. Geochem.*, 26, 1065–1076.

Nansheng, Q., Zhihuan, Z. and Ershe, X. (2008) Geothermal regime and Jurassic source rock maturity of the Junggar Basin, northwest China. *J. Asian Earth Sci.*, 31, 464–478.

Pallasser, R. J. (2000) Recognising biodegradation in gas/oil accumulations through the $\delta^{13}C$ compositions of gas components. *Org. Geochem.*, 31, 1363–1373.

Schoell, M. (1983) Genetic characterization of natural gases. *AAPG Bull.*, 67, 2225–2238.

Valyaev, B.M., Grinchenko, Y.I., Erokhin, V.E., Prokhorov, V.S., Titkov, G.A. (1985) Isotopic composition of gases from mud volcanoes. *Lithol. Miner. Resour.*, 20, 62–75.

Waseda, A., Iwano, H. (2008). Characterization of natural gases in Japan based on molecular and carbon isotope compositions. *Geofluids*, 8, 286–292.

Chapter 3

Petroleum Hydrocarbon Biodegradability in Soil – Implications for Bioremediation

Snežana Maletić, Božo Dalmacija and Srđan Rončević

Additional information is available at the end of the chapter

http://dx.doi.org/10.5772/50108

1. Introduction

The development of human civilization throughout history has led to growing disruption of the natural balance and the occurrence of different types of pollution. The world depends on oil, and the use of oil as fuel has led to intensive economic development worldwide. The great need for this energy source has led to the gradual exhaustion of natural oil reserves. However, mankind will witness the results of oil consumption for centuries after its cessation. Environmental pollution with petroleum and petrochemical products has been recognized as a significant and serious problem (Alexander, 1995, 2000). Most components of oil are toxic to humans and wildlife in general, as it is easy to incorporate into the food chain. This fact has increased scientific interest in examining the distribution, fate and behaviour of oil and its derivatives in the environment (Alexander, 1995, 2000; Semple et al., 2001, 2003; Stroud et al., 2007, 2009). Oil spills in the environment cause long-term damage to aquatic and soil ecosystems, human health and natural resources.

Petroleum oil spills tend to be associated with offshore oil rigs and tankers in marine-related accidents. In contrast, land oil spills often go unnoticed by everyone except environmentalists, yet land oil spills contribute to the pollution of our water supply and soil. Typical sources of land oil spills include accidents as well as oil from vehicles on the road.

Characterization of spilled oil and its derivatives is very important in order to predict the behaviour of oil and its long-term effects on the environment, and in order to select the proper cleaning methods. The potential danger which petroleum hydrocarbons pose to humans and the environment makes testing and characterization of the biodegradation and biotransformation processes of hydrocarbons in contaminated soil necessary in order to develop bioremediation techniques for cleaning such soils to levels that ensures its safe disposal or reuse. Biodegradation is the metabolic ability of microorganisms to transform or mineralize organic contaminants into less harmful, non-hazardous substances, which are then integrated into natural biogeochemical cycles. Petroleum hydrocarbon biodegradability in soil is influenced by complex arrays of factors, such as nutrients, oxygen, pH value, composition, concentration and bioavailability of the contaminants, and the soil's chemical and physical characteristics.

Bioremediation is considered a non-destructive, cost-effective, and sometimes logistically favourable cleanup technology, which attempts to accelerate the naturally occurring biodegradation of contaminants through the optimization of limiting conditions. In order to choose the appropriate bioremediation strategy it is extremely important to investigate and understand all factors which affect biodegradation efficiency. In order to better explain those factors, 4 examples of bioremediation studies (conducted 1 year, 5 years and 8 years after contamination) on soil which was directly contaminated with various petroleum products and their combustion products, are described along with their similarities and differences.

2. Bioremediation

Bioremediation can be briefly defined as the use of biological agents, such as bacteria, fungi, or green plants (phytoremediation), to remove or neutralize hazardous substances in polluted soil or water. Bacteria and fungi generally work by breaking down contaminants such as petroleum into less harmful substances. Plants can be used to aerate polluted soil and stimulate microbial action. They can also absorb contaminants such as salts and metals into their tissues, which are then harvested and disposed of. Bioremediation is a complex process, with biological degradation taking place in the cells of microorganisms which absorb pollutants, where if they have specific enzymes, the degradation of pollutants and their corresponding metabolites will take place. Hydrocarbons from oil are used as a source of nutrients and energy for microorganism growth, and at the same time, microorganisms decompose them to naphthenic acids, alcohols, phenols, hydroperoxides, carbonyl compounds, esters, and eventually to carbon dioxide and water (Eglinnton, 1975; Marković et al., 1996).

Bioremediation is considered a non-destructive, cost- and treatment-effective and sometimes logistically favourable cleanup technology, which attempts to accelerate the naturally occurring biodegradation of contaminants through the optimization of limiting conditions. Bioremediation is an option that offers the possibility to destroy or render harmless various contaminants using natural biological activity. As such, it uses relatively low-cost, low-technology techniques, which generally have a high public acceptance and can often be carried out on site (Alexander, 1995). It will not always be suitable, however, as the range of contaminants on which it is effective is limited, the time scales involved are relatively long, and the residual contaminant levels achievable may not always be appropriate (Maletić et al., 2009; Rončević et al., 2005).

Bioremediation can be divided into two basic types: (1) natural attenuation, which can be applied when the natural conditions are suitable for the performance of bioremediation without human intervention, and (2) engineered bioremediation, which is used when is necessary to add substances that stimulate microorganisms. The first one is more attractive because of its low cost, minimum of maintenance and minimal environmental impact. Still, this technology is applicable only in cases when the natural level of biodegradation is higher than the degree of pollution migration. Nevertheless, this technology is more often used as a supplement to the other technologies, or after finished engineered bioremediation in order to prevent migration of pollution from the treated area. Engineered bioremediation is faster than natural attenuation because it includes microbial degradation stimulation, by controlling the concentrations of pollution, oxygen, nutrients, moisture, pH, temperature, etc. (Rahman et al., 2003; Yerushalmi et al., 2003). Engineered bioremediation is applied when it is essential to carry out cleaning in a short time or when the pollution is very rapidly expanding. Its application reduces the costs due to the shorter treatment of land and lower number of sampling and analysis, and it is important for political and psychological needs when the community is exposed to pollution. Engineered bioremediation can be divided in two main groups (1) in situ and (2) ex situ bioremediation techniques, with the most applicable of these and their main characteristics given in Tables 1 and 2. In situ techniques are generally the most desirable options due to lower cost and fewer disturbances since they provide treatment in place, avoiding excavation and transport of contaminants (Vidali, 2001). In situ techniques are limited by the depth of the soil that can be effectively treated. In contrast, ex situ techniques involve the excavation or removal of contaminated soil from the ground.

3. Hydrocarbon biodegradation mechanisms and products

Biodegradation is the process by which microorganisms transform or mineralize organic contaminants, through metabolic or enzymatic processes, into less harmful, non-hazardous substances, which are then integrated into natural biogeochemical cycles. Organic material can be degraded by two biodegradation mechanisms: (1) aerobically, with oxygen, or (2) anaerobically, without oxygen.

Anaerobic processes are conducted by anaerobic microorganisms and this pathway of biodegradation is very slow. Originally thought to contribute marginally to overall biodegradation, anaerobic biodegradation mechanisms have been gaining more attention in recent years due to increased information regarding contaminant site conditions and rapid oxygen depletion (Burland & Edwards, 1999). Anaerobic biodegradation follows different biochemical pathways dependent on the electron acceptor utilized by the microorganism. Petroleum-based contaminants have been shown to degrade under various anaerobic conditions, including nitrate reduction, sulphate reduction, ferric iron reduction, manganese reduction and methanogenic conditions. The metabolic pathways behind anaerobic alkane biodegradation are not well understood. Most of the reports related to the anaerobic mineralization of aliphatic hydrocarbons are studies with pure cultures or enrichment cultures in laboratory scale experiments. Hence, the significance of these results in the environment e.g. in contaminated soils and sediments, is not yet known and evidence for the anaerobic degradation of alkanes in environmental samples has been reported in only a few cases (Salminen, 2004).

The most rapid and complete degradation of the majority of organic pollutants is brought about under aerobic conditions. The initial intracellular attack of organic pollutants is an oxidative process, and the activation and the incorporation of oxygen is the enzymatic key reaction catalyzed by oxygenases and peroxidases. Degradation pathways convert organic pollutants step by step into intermediates of the central intermediary metabolism, for example, the tricarboxylic acid cycle. Biosynthesis of cell biomass occurs from the central precursor metabolites, for example, acetyl-CoA, succinate, pyruvate. Sugars required for various biosyntheses and growth are synthesized by gluconeogenesis. The degradation of petroleum hydrocarbons can be mediated by specific enzyme systems. Other mechanisms involved are (1) attachment of microbial cells to the substrates and (2) production of biosurfactants. The uptake mechanism linked to the attachment of cell to oil droplet is still unknown but the production of biosurfactants has been well studied (Nilanjana & Chandran, 2011).

Technique / Definition	Advantages	Disadvantages	Applicability
Biosparging - Involves the injection of air under pressure below the water table to increase groundwater oxygen concentrations and enhance the rate of biological degradation of contaminants by naturally occurring bacteria (Baker & Moor, 2000; Khan et al., 2004).	Equipment is readily available and easy to install, little disturbance to site operations, treatment times from 6 months to 2 years, low injection rates reduce the need for vapour capture and treatment.	Can only be used in areas where air sparging is suitable, complex chemical, physical and biological processes are not well understood potential for the migration of contaminants.	Most types of petroleum contaminated sites, but it is least effective on heavy petroleum because of the length of time required.
Bioventing - injection of air into the contaminated media at a rate designed to maximize in situ biodegradation and minimize or eliminate the off-gassing of volatilized contaminants to the atmosphere (Khan et al., 2004).	Equipment is readily available and easy to install, short treatment times, from 6 months to 2 years. easy to combine with other technologies, may not require off-gas treatment.	High concentrations of contaminants may be toxic to organisms. cannot always reach low cleanup limits. is effective only in unsaturated soils; other methods are needed for the saturated zone.	Mid-weight petroleum products like diesel.
Phytoremediation - application of green plants to remove pollutants and other harmful components from the environment (Joner et al., 2006).	Cost-effective for large areas, no impact on the environment, formation of secondary waste is minimal, post-treatment soil can remain in the treated area and can be used in agriculture, uses solar energy no formation of toxic compounds.	Longer period required than one growing season, climate and hydro-logical conditions such may limit plant growth and the plant species that can be used, pollutants can enter the food chain, requires special disposal of plants.	Heavy metals, radionuclides, chlorinated solvents, petroleum hydrocarbons, insecticides, explosives and surfactants.

Table 1. The most applicable **in situ** techniques and their main characteristics

Technique / Definition	Advantages	Disadvantages	Applicability
Landfarming - spreading of contaminated soils in a thin layer on the ground surface of a treatment site and stimulating aerobic microbial activity within the soils through aeration and addition of nutrients, minerals, and water (Hejazi et al., 2003; Khan et al., 2004).	The most cost effective, takes less time and money to remediate, leads to complete destruction of pollutants, suitable for treating large volumes of contaminated soil.	Large amount of land required, VOC must be pre-treated not efficient for the heavy components of petroleum, possibility of contamination migration into the environment, difficult to expect a reduction in the concentration of pollutants greater than 95%.	Volatile organic compounds, gasoline, heating and lubricating oil, diesel oil, PAH etc.
Biopile A hybrid of landfarming and composting - engineered cells are constructed as aerated composted piles (Jorgensen et al., 2000).	May be constructed to suit a variety of terrain conditions, the treatment time - 6 months to 2 years, advantages over landfarming: takes up less space, possibility of aeration, VOC control is possible.	Not efficient for the heavy components of petroleum, possibility of contamination migration into the environment, difficult to expect a pollutants concentration reduction > 95%.	Petroleum products, non-halogenated and halogenated VOC and SVOC, PAH.
Composting - combining contaminated soil with non-hazardous organic materials which support the development of a rich microbial population and elevated temperature for composting (Semple et al., 2001).	Cost effective, takes less time and money to remediate leads to complete destruction of pollutants, suitable for treating large volumes of contaminated soil.	VOC must be pre-treated possibility of contamination migration into the environment, composting/compost processes to "lock up"' pollutants, the long-term stability of such "stabilized" matrices is uncertain.	Petroleum products, non-halogenated and halogenated VOC and SVOC, PAH, PCB and explosives, pesticides.
Bioslurry systems - the soil is treated in a controlled bioreactor where the slurry is mixed to keep the solids suspended and microorganisms in contact with the contaminants (Nano &Rota, 2003).	Control of temperature, moisture, pH, oxygen, nutrients, VOC emission, addition of surfactants, addition of micro-organisms, monitoring of reaction conditions.	Non-homogeneous and clayey soils can handling problems, free product removal is necessary, expensive soil dewatering after treatment, disposal method is needed for wastewater, extensive site and contaminant investigation.	Petroleum products, non-halogenated and halogenated VOC and SVOC, PAH, PCB and explosives.

Table 2. The most applicable **ex situ** techniques and their main characteristics

4. Bioremediation process kinetics

Bioremediation processes are time consuming and as a consequence, many studies have addressed the determination of bioremediation process kinetics. The kinetics for modelling the bioremediation of contaminated soils can be extremely complicated. This is largely due to the fact that the primary function of microbial metabolism is not for the remediation of environmental contaminants. Instead the primary metabolic function, whether bacterial or fungal in nature, is to grow and sustain more of the microorganism. Because of the involvement of adverse factors and the complexity of the process, it is not possible to predict the duration of bioremediation. Therefore, the formulation of a kinetic model must start with the active biomass and factors, such as supplemental nutrients and oxygen source that are necessary for subsequent biomass growth (Maletić et al., 2009; Rončević et al., 2005).

Studies of the kinetics of the bioremediation process proceed in two directions: (1) the first is concerned with factors influencing the amount of transformed compounds with time, and (2) the other approach seeks the types of curves describing the transformation and determines which of them fits the degradation of the given compounds by the microbiologic culture in the laboratory microcosm and sometimes, in the field.

Determinations based on the literature data for values of the degradation degree are useful but less exact, because they do not take into account all the specific characteristics of the soil such as temperature, moisture, and—most often—the adaptation of bacteria to the specific contaminants. A literature survey has shown that studies of biodegradation kinetics in the natural environment are often empiric, reflecting only a basic level of knowledge about the microbiologic population and its activity in a given environment. One such example of the empirical approach is the simple model:

$$\frac{dC}{dt} = kC^n \tag{1}$$

where C is the concentration of the substrate, t is time, and k is the degradation rate constant of the compound and n is a fitting parameter (most often taken to be unity) (Wethasinghe et al., 2006). Using this model, one can fit the curve of substrate removal by varying n and k until a satisfactory fit is obtained. It is evident from this equation that the rate is proportional to the exponent of substrate concentration. Researchers involved in kinetic studies do not always report whether the model they used was based on theory or experience and whether the constants in the equation have a physical meaning or if they just serve as fitting parameters (Maletić et al., 2009; Rončević et al., 2005).

With the complex array of factors that influence the biodegradation of hydrocarbons noted previously, it is not realistic to expect a simple kinetic model to provide precise and accurate descriptions of concentrations during different seasons and in different environments. The results of short-term degradation experiments are sometimes presented with the implicit assumption of zero-order kinetics (i.e., degradation in mass per unit time or in turnover time). However, short-term degradation experiments may not be adequate to discern the appropriate kinetics. In experiments with a number of samples taken during a length of time sufficient for considerable biodegradation to take place, the concentration of hydrocarbons with time is better described by first-order kinetics, eq. 2 (Collina et al., 2005; Grossi et al., 2002; Hohener et al., 2003; Pala et al., 2006; Rončević et al., 2005).

First order kinetics, such as the well known Michaelis-Menton kinetic model, is the most often used equation for the representation of degradation kinetics (Collina et al., 2005; Grossi et al., 2002; Hohener et al., 2003; Pala et al., 2006; Pollard et al., 2008). First order kinetics enables the prediction of hydrocarbon concentrations at any time from biodegradation half-times. If the optimal conditions are established, remediation time depends on the biodegradation half-time, initial hydrocarbon concentration in the polluted soil, and the end point concentration which needs to be achieved. Many researchers assume first order kinetics because of the easier presentation and data analysis, simplicity of graphical presentation, and the easier prediction of concentration once half-life has been determined [26, 28]. This approach is least reliable at very high and very low levels of contaminants. Taking the same initial values of concentration, different kinetic models will give significantly different final amounts of unreacted compound (Maletić et al., 2009; Rončević, 2002; Rončević et al., 2005):

$$C = C_0 e^{-kt} \text{ (or } \ln C = \ln C_0 - kt \text{)} \quad (2)$$

where C - concentration of hydrocarbons (g kg^{-1}), t - time of removal (day), C_0 - initial concentration of hydrocarbons (g kg^{-1}), and k - rate constant of the change in the hydrocarbon content of the soil (day^{-1}).

In the simple model, depending on the nature of the substrate and experimental conditions, various investigators obtain different values for the rate constant of substrate degradation: for n-alkanes, 0.14 to 0.61 day^{-1}; for crude oil, 0.0051 to 0.0074 day^{-1}; and for PAHs, 0.01 to 0.14 day^{-1} (Roncević et al., 2005). Reported rates for the degradation of hydrocarbon compounds under field or field-simulated conditions differ by up to two orders of magnitude. The selection of appropriate kinetics and rate constants is essential for accurate predictions or reconstructions of the concentrations of hydrocarbons with time in soil after a spill.

A more reliable prediction of pollution biodegradation can be obtained from more complex models such as the BIOPLUME II model (BIOPLUME is a two-dimensional computer model that simulates the transport of dissolved hydrocarbons under the influence of oxygen-limited biodegradation). Additionally, in recent years, the state of the art in modelling technology allows for even more reliable prediction using the 3D software MODFLOW, which is available in several versions: MODFLOW, MODPATH, MT3D, RT3D and MODFLOW-SURFACT).

For the ex-situ treatment of soil, remediation time generally does not depend on the transport of nutrients and oxygen and can be roughly determined from the degree of degradation, determined in laboratory tests of samples taken from the field. The following factors often interfere with a simple extrapolation of the kinetics described above in natural conditions:

1. Different barriers may limit or prevent contact between microbial cells and their organic substrates. Many organic molecules sorb to clay or soil humus or sediment, and the kinetics of the decomposition of sorbed substrate can be completely different from that of the same compound free in solution.
2. The presence of other organic molecules, which can be metabolized by biodegrading species can reduce or increase the consumption of the examined compounds.
3. Application of inorganic nutrients, oxygen, or growth factors, can affect the speed of transformation and then the process will be governed by diffusion of nutrients or the speed of their formation or regeneration of the other residents of the community.

4. Many species can metabolize the same organic compounds simultaneously.
5. Protozoa or possible species that parasitize on the biodegrading population can manage growth, population size or activity responsible for biodegradation.
6. Many synthetic chemicals have insufficient solubility in water, and the kinetics of their transformation can be completely different from compounds in the aqueous phase.
7. Cells of the active population may be sorbed or can develop microcolonies, and kinetics of sorbed or microcolonies is still unresolved.
8. Many organic compounds disappear only after a period of acclimatization, and there is no method that can predict the length of this period or the expected percentage of time between the occurrence of compounds and their total destruction.

4.1. Bioremediation study – Our experiences

In order to close this issue for readers, experience from four different bioremediation treatments of petroleum contaminated soil are given as examples (Fig. 1, Fig. 2). Thus, as a consequence of the accidental oil spill in the Novi Sad Oil Refinery (Serbia) in 1999, soil was directly contaminated with various petroleum products (gasoline, crude oil, kerosene, diesel fuel, black oil, etc.) and products of their combustion from frequent fires. Bioremediation studies on this soil were conducted 1 year after contamination (Rončević, 2002; Rončević et al., 2005), after 5 years (Rončević, 2007) and after 8 years (Maletić et al., 2009; Maletić, 2010; Maletić at al., 2011), and the bioremediation kinetics which were determined are compared here and discussed.

The obtained data from these four studies show changes and differences in the bioremediation kinetic rate, depending on the applied technology and stage of weathering (Fig 3.).

In study 1, bioremediation was carried out on a relatively freshly petroleum contaminated soil (one year after contamination), with a start concentration three times greater than in the other case studies, and the % of removed hydrocarbon is the highest. A slight difference was noticed between the two approaches applied (reactor with continuous and discontinuous flow). Namely, in the reactor with discontinuous flow, the hydrocarbon biodegradation rate in the aerobic part of the reactor was lower, indicating that the aerobic bioremediation conditions are favourable for this type of oil contaminated soil. Generally, satisfactory hydrocarbon degradation and removal rates were established by this technology.

As explained above, in study 2, the initial hydrocarbons concentration is three times lower, due to the different environmental conditions to which this soil was exposed during 5 years of weathering. Three varieties of in situ bioremediation technology were applied (Fig. 1). The first two used in-situ biostimulation feeding with aerated water and magnesium peroxide, and did not provide satisfactory results. The biodegradation kinetic rate constant could not be calculated, since no removal of hydrocarbons was observed during the bioremediation.

With the third variation, which used in situ biostimulation with ex situ biologically treated groundwater, the situation was changed drastically. Hydrocarbon content decreased rapidly, by about 60% in 232 days. Even so, the biodegradation kinetic rate constant is twice as low as the rate constant in study 1. This is probably because the degradation of the easily removable hydrocarbon fraction from the soil already occurred during the weathering process. Thus, only the heavier and less degradable fractions remained in the soil.

Bioremediation study 1 – Laboratory trial bioremediation

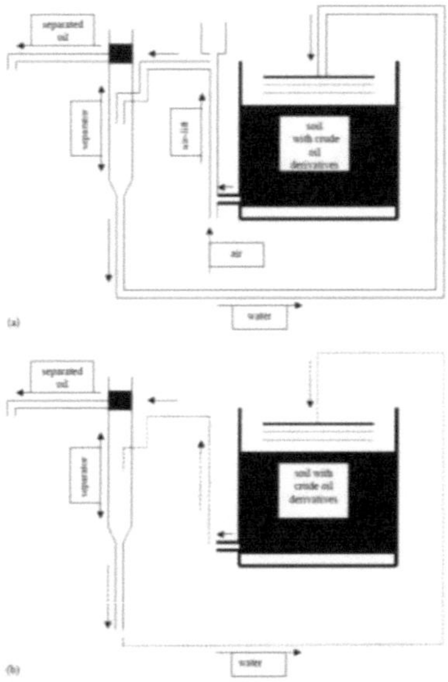

- Two samples of 170 and 180 kg were introduced into separate reactors which were filled to a height of 20 cm with water sampled from the piezometers from the refinery area. Experiment duration was 325 days.
- Each reactor was reinoculated daily by replacing 250 ml of the water phase with 250 ml of a suspension of adapted bacteria.
- One reactor (a) had continuous circulation of the water phase, with the aid of an air lift, at a flow rate of approximately 7 l day^{-1}.
- In the second (b), circulation of the water phase was carried out over a short period once a day to give a flow rate of 0.5 l day^{-1}.
- After percolating through the soil, the water phase was passed through a separator in which water insoluble components (free crude oil plus oil derivatives) were separated out by gravity. The separated-oily layer was removed periodically and fed into a third bioreactor that was used to prepare the adapted microbial suspension.

Bioremediation study 2 – Simulation of in situ bioremediation in a laboratory bioreactor

- Cylinder reactor, length 3.2 m and 0.8 m in diameter, with 4 piezometers placed in the soil.
- A layer of sand 10-15 cm thick was first placed in the reactor, then a layer of soil polluted with oil derivatives (thickness of 45-50 cm, 1150 kg of soil). 1 m^3 of groundwater from the site was added.
- 3 versions of the technical bioremediation were performed:
  - I - in-situ biostimulation feeding with aerated water - 2.7 dm^3 water was discharged into the aerator, where it was saturated with the maximum amount of oxygen, and poured over the surface of the soil at the beginning of the reactor. 306 days, changeable water flow 1.8-22 x10^{-7} m/s.
  - II - in-situ biostimulation with magnesium peroxide, the fourth piezometer was filled with magnesium peroxide, whose decomposition provides oxygen in the soil layer. 147 days, water flow 22 x10^{-7} m/s
  - III - in-situ biostimulation with ex situ biologically treated groundwater - water from the reactor was drained to a system consisting of three separators of the oil-free phase, a bioreactor, settler and sludge conditioner, and then recirculated over the surface of the soil at the beginning of the reactor. 232 days, changeable water flows 2.5-16 x10^{-7} m/s.

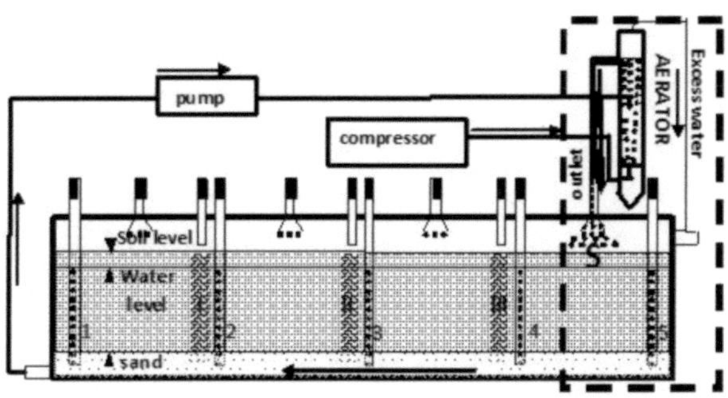

Figure 1. Experimental conditions for Bioremediation studies 1 and 2

Bioremediation study 3 – Biopile bioremediation

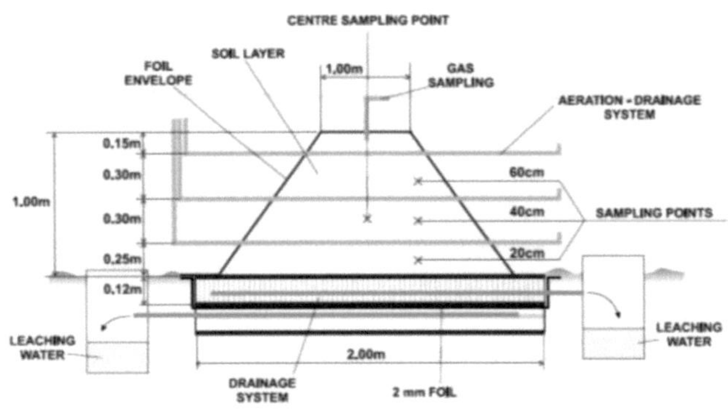

- The contaminated soil (2.7 m^3) was placed in a 2.2 m prismatic hole dug to a depth of 0.4 m, and covered with resistant polypropylene foil to prevent contamination spreading from the biopile.
- The layer of contaminated soil above the drainage system was in the form of a 1 m tall truncated pyramid composted with straw, and had a total volume of 2.7 m^3.
- To facilitate oxygen and water transport through the soil, the contaminated soil was composted with straw.
- At three different heights on the pyramid structure, perforated PVC aeration tubes were placed.
- To accelerate microbiological activity air was additionally piped through the biopile once a week.
- As well as stimulation of native microflora by soil aeration and irrigation, bioaugmentation was also carried out with microorganisms separated from the contaminated soil and cultivated in a laboratory bioreactor.
- The biopile was watered twice a week, and moisture was maintained at approximately 50-80% water holding capacity during the experiment. Leaching water from the biopile was collected in a separate reservoir and used for watering the biopile. Experiment duration 710 days.

Bioremediation study 4 – Landfarming bioremediation

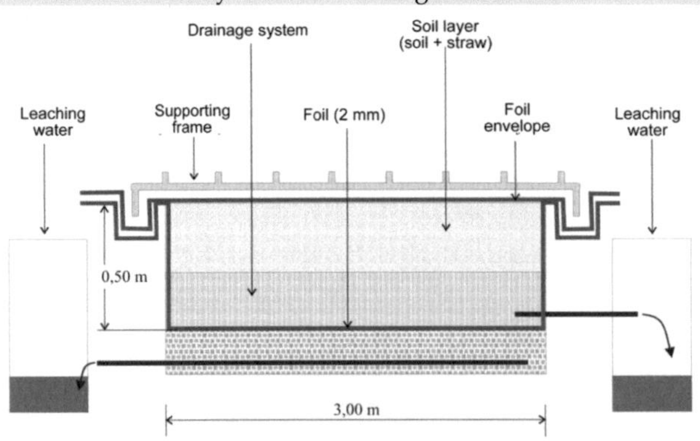

- The contaminated soil (2.7 m^3) was placed in a 3x3 m wide and 0.4 m deep prismatic hole, and covered with resistant polypropylene foil to prevent contamination spreading from the landfarm. Experiment duration 710 days.
- To facilitate oxygen and water transport through the soil, the soil was composted with straw.
- The landfarm was turned twice a month and watered twice a week; moisture was maintained at approximately 50-80% water holding capacity during the experiment.
- In addition to the stimulation of native microflora by soil aeration and irrigation, bioaugmentation was also carried out with microorganisms separated from the contaminated soil and cultivated in a laboratory bioreactor.
- Approximately 25 dm^3 of the inoculated water from the bioreactor was used together with leaching water for weathering the landfarm.

Figure 2. Experimental conditions for Bioremediation studies 3 and 4

Total petroleum hydrocarbon

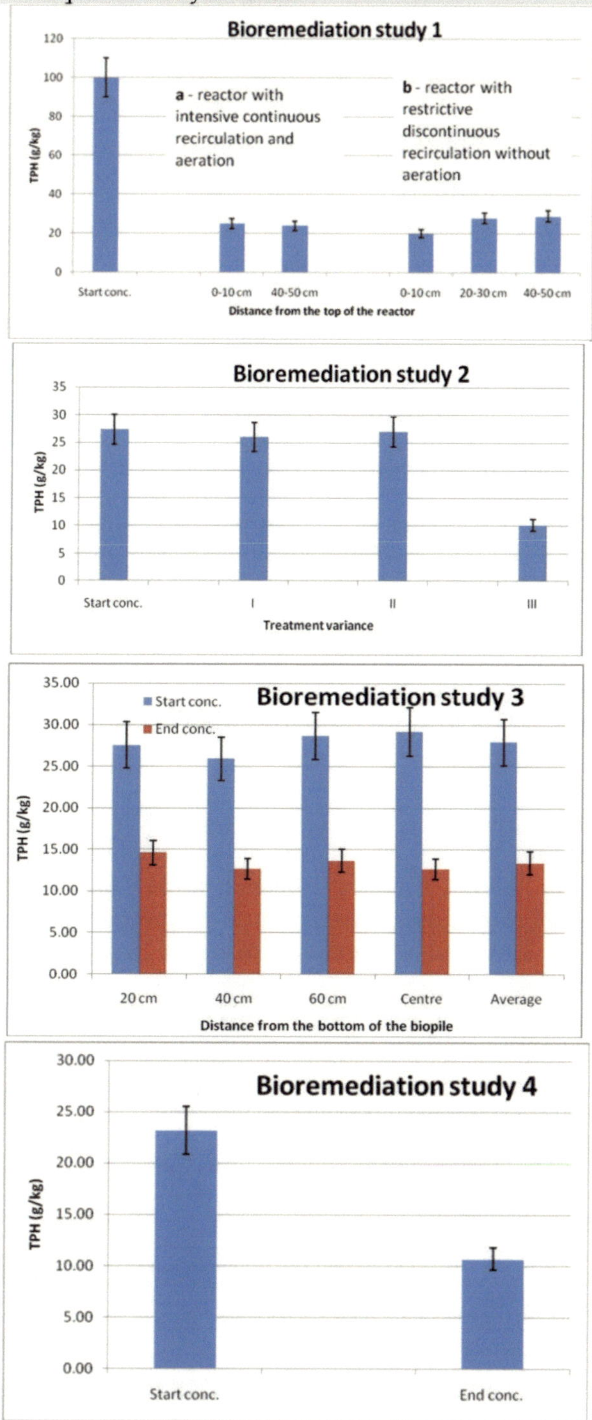

Kinetic parameters

Study	lnCo	k (day^{-1})	r
Case study 1a – 0-10 cm	4.5	0.0052	0.90
Case study 1a – 40-50 cm	4.5	0.0046	0.96
Case study 1b – 0-10 cm	4.6	0.0057	0.98
Case study 1b – 20-30 cm	4.6	0.0046	0.95
Case study 1b – 40-50 cm	4.5	0.0045	0.92
Case study 2 - I	-	-	-
Case study 2 - II	-	-	-
Case study 2 - III	3.3	0.0083	0.97
Case study 3 – 20 cm	3.2	0.00052	0.82
Case study 3 – 40 cm	3.2	0.00080	0.85
Case study 3 – 60 cm	3.3	0.00093	0.94
Case study 3 – centre	3.3	0.00078	0.90
Case study 3 – average	3.2	0.00077	0.96
Case study 4	3.1	0.00065	0.79

Co – start concentration
k – rate constant
r – correlation coefficient

Figure 3. Experimental results from the bioremediation studies 1-4

Studies 3 and 4 had similar hydrocarbon concentrations at the beginning of the experiment as study 2; even so, the biodegradation constant rate for both case studies is one order of magnitude lower than in study 2. The reason for this could be hydrocarbon complexation with the soil organic material and also its sorption and sequestration in the soil nanopores with further weathering of the oil contaminated soil (8 years). In this manner the hydrocarbons become recalcitrant and resistant to biodegradation. In study 3 (biopile) the hydrocarbon biodegradation removals were also monitored at different heights in the biopile.

Similarly to study 1, the lowest biodegradation rate constant was obtained for the lowest layer of the biopile, where the oxygen concentration is limited and anaerobic conditions developed. This confirms the facts from study 1 that aerobic degradation of hydrocarbons is the favourable degradation pathway. It is worth mentioning that in general, greater rate constants were obtained in the biopile than in the landfarming, indicating that the biopile is a better technology choice for bioremediation of this type of soil contamination.

5. Factors affecting oil hydrocarbon biodegradation processes

Successful implementation of bioremediation technologies on contaminated areas depends on the characteristics of the contaminated site and a complex system of many factors that affect the petroleum hydrocarbons biodegradation processes (Jain et al., 2011). The main factors which limit the overall biodegradation rate can be grouped as: soil characteristics, contaminant characteristics, bioavailability, microorganisms number and catabolism evolution (Alexander, 1995). In order to adopt and implement some bioremediation strategy it is extremely important to consider and understand those limiting factors.

5.1. Soil characteristics

Soil characteristics are especially important for successful hydrocarbon biodegradation, some of the main limiting factors are: soil texture, permeability, pH, water holding capacity, soil temperature, nutrient content and oxygen content. Soil texture affects permeability, water content and the bulk density of soil. Soil with low permeability (such as clays) hinders transportation and the distribution of water, nutrients and oxygen. To enable the bioremediation of such soil, it should be mixed with amendments or bulking materials (straw, sawdust etc.), as the bioremediation processes rely on microbial activity, and microorganisms require oxygen inorganic nutrients, water and optimal temperature and pH to support cell growth and sustain biodegradation (Alexander, 1995; Jain et al., 2011). The optimal conditions for microbial growth and hydrocarbon biodegradation are given in table 3.

Parameter	Microbial growth	HC biodegradation
Water holding capacity	25 -28	40-80
pH	5.5-8.8	6.5-8.0
Temperature (°C)	10-45	20-30
Oxygen (air-filled pore space)	10%	10-40%
C:N:P	100:10:1(0.5)	100:10:1(0.5)
Contaminants	Not too toxic	HC 5–10% of dry weight of soil
Heavy metals	<2000 ppm	<700 ppm

Table 3. Optimal conditions for microbial growth and hydrocarbon biodegradation

5.2. Contaminant characteristics

Petroleum hydrocarbons contain a complex mixture of compounds; all the components of petroleum do not degrade at the same rate. The rate by which microorganisms degrade hydrocarbons depends upon their chemical structure and concentration. Petroleum hydrocarbons can be categorized into four fractions: saturates, aromatics, resins and asphaltene. Of the various petroleum fractions, n-alkanes of intermediate length (C_{10}-C_{25}) are the preferred substrates for microorganisms and tend to be the most readily degradable, whereas shorter chain compounds are rather more toxic. Longer chain alkanes (C_{25}-C_{40}) are hydrophobic solids and consequently are difficult to degrade due to their poor water solubility and bioavailability, and branched chain alkanes and cycloalkanes are also degraded more slowly than the corresponding normal alkanes. Highly condensed aromatic and cycloparaffinic structures, tars, bitumen and asphaltic materials, have the highest boiling points and exhibit the greatest resistance to biodegradation. It has been suggested that the residual material from oil degradation is analogous to, and can even be regarded as, humic material (Balba et al., 1998; Loeher et al., 2001; Ivančev-Tumbas et al., 2004; Brassington et al., 2007; Stroud et al., 2007).

5.3. Bioavailability

Even if the optimal conditions for hydrocarbon biodegradation are provided at the field, it has been shown that a residual fraction of hydrocarbon remains undegraded. Namely, after its arrival in the soil, an organic contaminant may be lost by biodegradation, leaching or volatilization, or it may accumulate within the soil biota or be sequestered and complex within the soil's mineral and organic matter fractions. The rate at which hydrocarbon-degrading microorganisms can convert chemicals depends on the rate of transfer to the cell and the rate of uptake and metabolism by the microorganisms. It is controlled by a number of physical-chemical processes such as sorption/desorption, diffusion, and dissolution. (Brassington et al., 2007; Cuypers et al., 2002; Maletić et al., 2011; Semple et al., 2003). The mass transfer of a contaminant determines microbial bioavailability. The term "bioavailability" refers to the fraction of chemicals in soil that can be utilized or transformed by living organisms. The bioavailability of a compound is defined as the ratio of mass transfer and soil biota intrinsic activities. Most soil contaminants show biphasic behaviour, whereby in the initial phase of hydrocarbon biodegradation, the rate of removal is high and removal is primarily limited by microbial degradation kinetics. In the second phase, the rate of hydrocarbon removal is low and removal is generally limited by slow desorption. Altogether, the poorly bioavailable fraction of hydrocarbon contamination is formed by hydrocarbons which desorb slowly in the second phase of bioremediation (Loeher et al., 2001). The biodegradation of an oil-contaminated soil can also be seriously affected by the contamination time, due to weathering processes, which decrease the bioavailability of pollutants to microorganisms. Weathering refers to the results of biological, chemical and physical processes that can affect the type of hydrocarbons that remain in a soil (Maletić et al., 2011; Loeher et al., 2001; Semple et al.,2005). Those processes enhance the sorption of hydrophobic organic contaminants to the soil matrix, decreasing the rate and extent of biodegradation. Moreover, a weathered oil-contaminated soil normally contains a recalcitrant fraction of compounds composed basically of high molecular weight hydrocarbons, which cannot be degraded by indigenous microorganisms (Balba et al., 1998; Maletić et al., 2011; Loeher et al., 2001). In contrast, a recently oil-contaminated soil contains a higher amount of saturated and aliphatic compounds, which are the most susceptible to microbial degradation. However, the pollutant compounds in a recently contaminated soil are potentially more toxic to the native microorganisms, leading to a longer adaptation time (lag phase) before degradation of the pollutant and even to an inhibition of the biodegradation process (Margesin et al., 2000; Loeher et al., 2001; Petrović et al., 2008).

As was mentioned above, sequestration and weathering of organic contaminants in the soil reduces the bioavailability of organic compounds and results in non-degraded residues in the soil. Contaminants that have been weathered and sequestrated in soil are not available for biodegradation in soil, even though freshly added compounds are still biodegradable (Alexander, 1995). Sorption is a major factor preventing the complete bioremediation of hydrocarbons in soil. Slow sorption leads to the hydrocarbon fraction becoming resistant to desorption and increases its persistence within the soil organic matrix. The following hypotheses have been proposed as a explanation for weathering: (1) weathering results in a slow diffusion of the hydrocarbon fraction in the solid fraction of the organic matter in the soil; (2) the contaminant slowly diffuses through the soil and becomes sorbed and trapped in the soil nano-and micropores (Semple et al., 2003; Trinidade et al., 2005).

5.4. Microorganisms number and catabolism evolution

The ability of the soil's microbial community to degrade hydrocarbons depends on the microbes number and its catabolic activity. Microorganisms can be isolated from almost all environmental conditions. Soil microflora contain numbers of different microorganisms including bacteria, algae, fungi, protozoa and actinomycetes, which have a diverse capacity for attacking hydrocarbons. The main factors which affect the rate of microbial decomposition of hydrocarbons are: the availability of the contaminants to the microorganisms that have the catabolic ability to degrade them; the numbers of degrading microorganisms present in the soil; the activity of degrading microorganisms, and the molecular structure of the contaminant (Semple et al., 2003). The soil microorganisms number is usually in the range 10^4 to 10^7 CFU, for successful biodegradation this number should not be lower than 10^3 per gram of soil. Microorganism numbers lower than 10^3 CFU per gram of soil indicate the presence of toxic concentrations of organic or inorganic contaminants (Margesin et al., 2000; Petrović et al., 2008). The activity of soil microflora can be controlled by the factors discussed above - pH, temperature, nutrients, oxygen etc. For successful biodegradation, it is also necessary that the microorganisms can develop catabolic activity, by the following activities: induction of specific enzymes, development of new metabolic capabilities through genetic changes, and selective enrichment of organisms able to transform the target contaminant (Margesin et al., 2000, Semple et al., 2003).

5.5. Bioremediation study – Our experiences

With the aim of better understanding the factors which affect hydrocarbon biodegradation, results from the bioremediation studies described above are also given here, along with a comparison and discussion of changes in hydrocarbon composition and bioavailability over the years (Ivančev-Tumbas et al., 2004; Maletić, 2010; Maletić at al., 2011; Rončević, 2002; Rončević, 2007). The compounds detected by GC-MS analysis of extracts of the various soil samples taken at the start and end of bioremediation studies are given in Fig. 1, with only the main compounds from the hit lists of the probability-based matching (PBM ≥ 60%) search given.

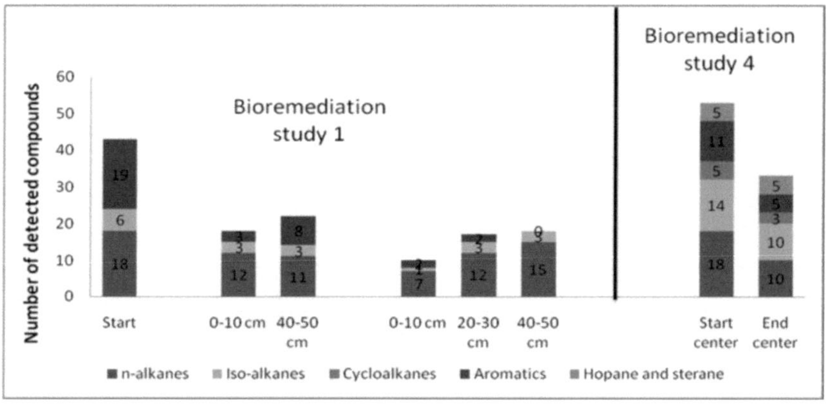

Figure 4. GC-MS SCAN qualitative analysis of soil samples

The data reflect the fact that the soil used in this investigation was sampled from the dumping area of a refinery where the initial pollutants were of very diverse composition, i.e. a mixture of crude oil, mazut, diesel, middle distillates, heavy distillates, kerosene, etc. The untreated soil samples contained a large variety of straight-chain hydrocarbons and their methyl derivatives (those with both even and odd numbers of C atoms), many of which persisted during the treatment. However, if we compare the untreated soil samples at the start of study 1 (1 year after contamination), and study 4 (8 years after contamination) the difference is significant. Namely, in study 1, the soil mostly contains n-alkanes and derivates of aromatic hydrocarbons, and few compounds of iso-alkanes, whereas the soil in study 4 contains mostly n-alkanes and iso-alkanes, with only a few aromatics derivatives detected, with PBM<50%, and few cycloalkanes. The fact that mainly substituted polycyclic aromatic hydrocarbons were not detected in the weathered soil samples (study 4), shows their lower persistence than alkanes. Additionally, the greater number of iso-alkanes in weathered soil indicates their persistence. The cycloalkanes detected represent one of the main hydrocarbon residual fractions in weathered contaminated soil.

In both studies, at the end of the experiment, the number of detected compounds is significantly reduced. In study 1, the aromatic hydrocarbons were almost completely removed in both reactors, while the number of n-alkanes detected was reduced, but they are still present in significant numbers in the soil at the end. This is a because the aromatics have lower persistence than n-alkanes, but is also due to the higher n-alkanes concentration at the beginning. It is worth mentioning that in the reactor with continuous flow (aerobic), the number of removed n-alkanes is almost the same, while in the reactor with discontinuous flows (partially anaerobic), the number of removed n-alkanes progressively reduced with depth, as a consequence of the lack of oxygen for microbial degradation, indicating that for this type of hydrocarbon, aerobic conditions are favourable. No such observation was noticed for aromatics. In study 4, only 3 n-alkanes compounds were detected at the end, also the number of poorly degradable iso-alkanes was also significantly reduced; this could be consequence of the lack of more degradable substrate which was probably removed during the weathering process.

Although the number of detected compounds (Fig. 4) and TPH concentration (Fig. 3) at the end was significant, the bioremediation rate was too slow to suggest that further bioremediation was possible. With the aim of investigating whether the lack of further hydrocarbon biodegradation was a consequence of the absence of the bioavailable hydrocarbon fraction for microbial degradation, the accumulation of toxic hydrocarbon degradation by-products, or the high concentration of hydrocarbons, a laboratory trial on the soil from study 4 was conducted (Maletić, 2010; Maletić et al., 2011). Study 4 was carried out for almost 2 years, however, after about one year, the biodegradation process slowed down significantly; at that point, some of the soil from study 4 was taken for the laboratory trial. The laboratory trials aimed in two directions: (1) bioavailability and (2) biodegradability investigation. Additionally, in order to test the impact of concentration, chemical composition and weathering on the biodegradation processes, the same tests were conducted on soil freshly contaminated by crude oil and diesel oil [36]. The bioavailability test was done by extraction of hydrocarbon contaminated soil with Tween 80. Table 4 shows the main results obtained from this test. To test whether high concentration or the accumulation of toxic by-products was the reason for the lack of biodegradation, the same soil sample was diluted with clean soil and then subjected to biodegradation under laboratory conditions (48 days). To ensure the process was not limited by other factors, the optimal conditions was provided, with respect to pH, temperature, water holding capacity, nutrients and oxygen content. The biodegradation process was monitored by measuring daily CO_2 production and TPH concentrations at the beginning and at the end of the experiment (Table 5).

The obtained results show that only 33% of the total amount of TPH is bioavailable in the weathered oil contaminated soil (soil taken from study 4). In the freshly contaminated soil, the bioavailable TPH fraction was three times larger, clearly indicating that in the weathered contaminated soil, the hydrocarbon is highly sequestrated in the soil pores and complexed with soil organic matter. As a result of these processes, petroleum hydrocarbons become resistant and unavailable for biodegradation.

Parameter	Type of the soil contaminant		
	Weathered oil	Crude oil	Diesel oil
TPH g/kg at the beginning	12 (±1.2)	26 (±2.6)	28 (±2.8)
TPH g/kg residual after Tween extraction	8 (±0.8)	3.6 (±0.4)	1.2 (±0.1)
%removed by Tween extraction	33	86	96

Table 4. Laboratory bioavailability trial results

The biodegradation study showed there was little difference between the respiration of the original and diluted samples of weathered oil contaminated soil (Table 5.). The evolved CO_2 from those samples could originate from basal microbial respiration and from the very slow degradation of poorly biodegradable hydrocarbon compounds. This is confirmed by the removed amount of TPH in the samples. In contrast, in the freshly contaminated soil, respiration and the amount of TPH removed both strongly depended on the TPH concentration and origin. Thus, the highest quantity of evolved CO_2 was produced by the soil contaminated with diesel oil (16 mg TPH/g), with the sample contaminated with crude oil (13 mg TPH/g) producing a slightly lower cumulative quantity of evolved CO_2. The sample which contained the highest TPH concentration in the soils contaminated with diesel or crude oil had a lower respiration, which is a consequence of the high level of soluble hydrocarbons and the possible generation of toxic biodegradation products which can be toxic to the microorganisms present. Likewise, the sample with soil contaminated with the highest amount of diesel oil produced the second smallest amount of CO_2 in the range of diesel contaminated soils. Thus, the diesel oil contains mostly midrange alkanes which have varying solubility and can cause toxic effects. The smallest amount of evolved CO_2 was obtained for the samples with the lowest TPH concentrations of diesel and crude oil, where the biodegradable fraction was readily degraded. The amounts of TPH removed were in general agreement with the respiration rate, but less TPH was removed from the samples with crude oil contaminated soil. This could be due to the higher amounts of polar hydrocarbons (which are not included in the TPH fraction) in crude oil which can be degraded faster than the TPH.

From comparing the end TPH concentration in the biodegradation sample on the original weathered oil contaminated soil (Table 5), and the predicted bioavailable fraction (Table 4), it can be concluded that a small amount of bioavailable substrate remained at the end of the treatment. Nevertheless, it should be borne in mind that the bioavailability test was conducted at the beginning of the experiment, and that as well as the biodegradation processes during the experiment, the sorption and sequestration of hydrocarbons also took place. These processes reduced the bioavailable hydrocarbon fraction during the treatment. Additionally, it is worth mentioning that during the 2 years of bioremediation study 4, the TPH concentration was reduced by 53% (21% in the first year and 32% in the second year), indicating that all of the biodegradable TPH fraction was removed during the treatment.

From the above discussion it can be concluded that the lack of hydrocarbon biodegradation was due to highly sorbed and sequestrated hydrocarbons in the soil pores and soil organic matter as a consequence of weathering, and not due to high hydrocarbon concentrations or accumulation of toxic products in the soil. This soil is therefore not suitable for further bioremediation, and if further removal of hydrocarbons is required, other technologies must be applied.

Contaminated soil	TPH g/kg after dilution	Evolved CO_2 mg/g	g/kg removed TPH
Weathered oil contaminated oil	12 (±1.2) (original soil)	6.1 (±0.9)	2.2 (±0.2)
	4.9 (±0.5)	6.8 (±1.0)	1.3 (±0.1)
	3.8 (±0.4)	5.2 (±0.8)	1.5 (±0.2)
	2.3 (±0.2)	4.6 (±0.7)	0.76 (±0.1)
Crude oil contaminated soil	26 (±2.6)	15 (±2.2)	16 (±1.6)
	13 (±1.3)	20 (±3.0)	11 (±1.1)
	7.5 (±0.8)	11 (±1.7)	6.5 (±0.7)
	5.5 (±0.6)	5.3 (±0.8)	4.4 (±0.4)
Diesel oil contaminated soil	28 (±2.8)	14 (±2.2)	11 (±1.1)
	16 (±1.6)	23 (±3.4)	11 (±1.1)
	9.2 (±0.9)	17 (±2.6)	6.8 (±0.7)
	7.0 (±0.7)	7.9 (±1.2)	5.1 (±0.5)

Table 5. Laboratory biodegradability results

6. Conclusion

The cleaning up of petroleum hydrocarbons in the soil environment is a real world problem. Better understanding of the mechanisms and factors which affect biodegradation is of great ecological significance, since the choice of bioremediation strategy depends on it. Microbial degradation processes aid the elimination of spilled oil from the environment, together with various physical and chemical methods. This is possible because microorganisms have enzyme systems to degrade and utilize different hydrocarbons as a source of carbon and energy. Even if the optimal conditions for microbial degradation are provided, the extent of hydrocarbon removal is strongly affected by its bioavailability and stages of weathering. As a consequence, some fractions of hydrocarbons remain undegraded. This residual fraction of hydrocarbon in soil can represent an acceptable end point for bioremediation if (1) hydrocarbon biodegradation is too slow to allow further bioremediation, in which case other technologies must be applied; (2) those concentrations are unable to release from the soil and pose adverse effects to the environment and human health, like those presented in the given case studies. Such residual material from oil degradation is analogous to, and could even be regarded as, humic material. Its inert characteristics, insolubility and similarity to humic materials mean it is unlikely to be environmentally hazardous.

Author details

Snežana Maletić, Božo Dalmacija and Srđan Rončević
University of Novi Sad Faculty of Sciences, Department of Chemistry,
Biochemistry and Environmental Protection,
Republic of Serbia

Acknowledgement

This research was financed by the Ministry of Education and Science (Projects No. III43005 and TR37004) and the Novi Sad Oil Refinery.

7. References

Alexander, M. (1995). How toxic are toxic chemicals in soil? *Environmental Science and Technology*, Vol. 29, No. 11, pp. 2713–2717, ISSN 0013-936X.

Alexander, M. (2000). Aging, bioavailability, and overestimation of risk from environmental pollutants. *Environmental Science and Technology,* Vol. 34, No. 20,pp. 4259–4265, ISSN 0013-936X.

Baker, R.S. & Moore, A.T. (2000). Optimizing the effectiveness of in situ bioventing. *Pollution Engenering*, Vol. 32, No. 7, pp. 44–47, ISSN 0032-3640.

Balba, M.T.; Al-Awadhi, N. & Al-Daher, R. (1998). Bioremediation of oil-contaminated soil: microbiological methods for feasibility assessment and field evaluation. *Journal of Microbiological Methods*, Vol. 32, No. 2, pp. 155–164, ISSN 0167-7012.

Brassington, K.J.; Hough, R.L.; Paton, G.I.; Semple, K.T.; Risdon, G.C.; Crossley, J.; Hay, I.; Askari, K. & Pollard, S.J.T. (2007). Weathered Hydrocarbon Wastes: A Risk Management Primer. *Critical Reviews in Environmental Science and Technology,* Vol. 37, No. 3, pp. 199-232, ISSN 1064-3389.

Burland, S.M. & Edwards, E.A. (1999). Anaerobic benzene biodegradation linked to nitrate reduction. *Applied and Environmental Microbiology,* Vol. 65, No. 2, pp. 529-533, ISSN 0099-2240.

Collina, E.; Bestetti, G.; Di Gennaro, P.; Franzetti, A.; Gugliersi, F.; Lasagni, M. & Pitea, D. (2005). Naphthalene biodegradation kinetics in an aerobic slurry-phase bioreactor. *Environment International*, Vol. 31, No. 2, pp. 167– 171, ISSN 0160-4120.

Cuypers, C.; Pancras, T.; Grotenhuis, T. & Rulkens, W. (2002). The estimation of PAH bioavailability in contaminated sediments using hydroxypropyl- β-cyclodextrin and triton X-100 extraction techniques. *Chemosphere,* Vol. 46, No. 8, pp. 1235–1245, ISSN 0045-6535.

Eglinnton, G. (1975). *Environmental chemistry,* Vol. 1, Specialist periodical reports. The Chemical Society, Burlington House, ISBN 0851867553, London.

Grossi, V.; Massias, D.; Stora, G. & Bertrand, J.C. (2002). Exportation and degradation of acyclic petroleum hydrocarbons following simulated oil spill in bioturbated Mediterranean coastal sediments. *Chemosphere,* Vol. 48, No. 9, pp. 947–954, ISSN 0045-6535.

Hejazi, R.; Husain, T. & Khan, F.I. (2003). Landfarming operation in arid region-human health risk assessment. *Journal of Hazardous Materials,* Vol. 99, No. 3, pp. 287–302, ISSN 0304-3894.

Hohener, P.; Duwig, C.; Pasteris, G.; Kaufmann, K.; Dakhel, N. & Harms, H. (2003). Biodegradation of petroleum hydrocarbon vapours: Laboratory studies on rates and kinetics in unsaturated alluvial sand. *Journal of Contaminant Hydrology*, Vol. 66, No. 1-2, pp. 93–115, ISSN 0169-7722.

Ivančev-Tumbas, I.; Tričković, J.; Karlović, E.; Tamaš, Z.; Rončević, S.; Dalmacija, B.; Petrović, O. & Klašnja, M. (2004). GC/MS-SCAN to follow the fate of crude oil components in bioreactors set to remediate contaminated soil. *International Biodeterioration and Biodegradation*, Vol. 54, No. 4, pp. 311-318, ISSN 0964-8305.

Jain, P.K.; Gupta, V.K.; Guar, R.K.; Lowry, M.; Jaroli, D.P. & Chauhan, U.K. (2011). Bioremediation opf petroleum contaminated soil and water. *Research Journal of Environmental Toxicology*, Vol. 5, No. 1, pp. 1-26, ISSN 1819-3420.

Joner, E.J.; Leyval, C. & Colpaert, V.J. (2006). Ectomycorrhizas impede phytoremediation of polycyclic aromatic hydrocarbons (PAHs) both within and beyond the rhizosphere. *Environmental Pollution*, Vol. 142, No. 1, pp. 34-38, ISSN 0269-7491.

Jorgensen, K.S.; Puustinen, J. & Suortti, A.M. (2000). Bioremediation of petroleum hydrocarbon-contaminated soil by composting in biopiles. *Environmental Pollution*, Vol. 107, No. 2, pp. 245–254, ISSN 0269-7491.

Khan, I.F.; Husain, T. & Hejazi, R. (2004). An overview and analysis of site remediaion technologies. *Journal of Environmental Management*, Vol. 71, No. 2, pp. 95-122, ISSN 0301-4797.

Loeher, R.C.; Mc Millen, S.J. & Webster, M.T. (2001). Predictions of biotreatability and actual results: soils with petroleum hydrocarbons. *Pratice periodical of hazardous, toxic, and radioactive waste management*, Vol. 5, No. 2, pp. 78–87, ISSN 1090-025X.

Maletić, S. (2010) Characterisation of the biodegradability of petroleum hydrocarbons in soil and the bioremediation processes during treatment by biopiles and landfarming, *PhD Thesis*.

Maletić, S.; Dalmacija, B.; Rončević, S.; Agbaba, J. & Ugarčina Perović, S. (2011). Impact of hydrocarbon type, concentration and weathering on its biodegradability in soil. *Journal of Environmental Science and Health, Part A*, Vol. 46, No. 10, pp. 1042-1049, ISSN 1093-4529.

Maletić, S.; Dalmacija, B.; Rončević, S.; Agbaba, J. & Petrović, O, (2009). Degradation Kinetics of an Aged Hydrocarbon-Contaminated Soil. *Water Air Soil Pollution*, Vol. 202, No. 1-4, pp. 149-159, ISSN 0049-6979.

Margesin, R.; Zimmerbauer, A. & Schinner, F. (2000). Monitoring of bioremediation by soil biological activities. *Chemosphere*, Vol. 40, No. 4, pp. 339–346, ISSN 0045-6535.

Marković, D.A.; Đarmati, Š.A.; Gržetić, I.A. & Veselinović, D.S. (1996). *Fizičkohemijski osnovi zaštite životne sredine, Izvori zagađivanja, posledice i zaštita*. ISBN 86-81019-27-9 Univerzitet u Beogradu, Beograd.

Namkoonga, W.; Hwangb, E.Y.; Parka, J.S. & Choic, J.Y. (2002). Bioremediation of diesel-contaminated soil with composting. *Environmental Pollution*, Vol. 119, No. 1, pp. 23–31, ISSN 0269-7491.

Nano, A.G. & Rota, R. (2003).Combined slurry and solid –phase bioremediation of diesel contaminated soils. *Journal of Hazardous Materials*, Vol. 100, No. 1-3, pp. 79–94, ISSN 0304-3894.

Nilanjana, D. & Chandran, P. (2011). Microbial Degradation of Petroleum Hydrocarbon Contaminants: An Overview. *Biotechnology Research International*, Vol. 2011, pp. 1-13, ISSN 2090-3138.

Pala, D.M.; de Carvalho, D.D.; Pinto, J.C. & Sant'Anna, Jr G.L. (2006). A suitable model to describe bioremediation of a petroleum-contaminated soil. *International Biodeterioration and Biodegradation*, Vol. 58, No. 3-4, pp. 254-260, ISSN 0964-8305.

Petrović, O.; Knežević, P.; Marković, J. & Rončević, S. (2008). Screening method for detection of hydrocarbon-oxidizing bacteria in oil-contaminated water and soil specimens. *Journal of Microbiological Methods*, Vol. 74, No. 2-3, pp. 110–113, ISSN 0167-7012.

Pollard, S.J.T.; Hough, R.L.; Kim, K.; Bellarby, J.; Paton, G.I.; Semple, K.T. & Coulon, F. (2008). Fugacity modelling to predict the distribution of organic contaminants in the soil:oil matrix of constructed biopiles. *Chemosphere*, Vol. 71, No. 8, pp. 1432–1439, ISSN 0045-6535.

Rahman, K.S.M.; Rahman, T.J. Kourkoutas, Y.; Petsas, I.; Marchant, R. & Banat, I.M. (2003). Enhanced bioremediation of n–alkane in petroleum sludge using bacterial consortium amended with rhamnolipid and micronutrients, *Bioresource Technology*, Vol. 90, No. 2, pp. 159–168, ISSN 0960-8524.

Rončević, S. (2002). Kinetics of bioremediation processes in soil contaminated by oil and oil derivates, *MSc Thesis*.

Rončević, S. (2007). Characterisation of the bioremediation processes in soil and groundwater contaminated by oil and oil derivatives at the site of Ratno Ostrvo, *PhD Thesis*.

Rončević, S.; Dalmacija, B.; Ivančev-Tumbas, I.; Petrović, O.; Klašnja, M. & Agbaba, J. (2005). Kinetics of Degradation of Hydrocarbons in the Contaminated Soil Layer. *Archives of Environmental Contamination and Toxicology*, Vol. 49, No. 1, pp. 27-36, ISSN 0090-4341.

Salminen, J.M.; Tuomi, P.M.; Suortti, A. & Jørgensen, K.S. (2004). Potential for Aerobic and Anaerobic Biodegradation of Petroleum Hydrocarbons in Boreal Subsurface. *Biodegradation*, Vol. 15, No. 1, pp. 29-39, ISSN 0923-9820.

Semple, K.T.; Morris, A.W.J. & Paton, G.I. (2003). Bioavailability of hydrophobic organic contaminants in soils: fundamental concepts and techniques for analysis. European *Journal of Soil Science*, Vol. 54, pp. 809-818, ISSN 0022-4588.

Semple, K.T.; Reid, B.J. & Fermor, T.R. (2001). Impact of composting strategies on the treatment of soils contaminated with organic pollutants. *Environmental Pollution*, Vol. 112, No. 2, pp. 269-283, ISSN 0269-7491.

Stroud, J.L.; Paton, G.I. & Semple, K.T. (2007). Microbe–aliphatic hydrocarbon interactions in soil: implications for biodegradation and bioremediation. *Journal of Applied Microbiology*, Vol. 102, No. 5, pp. 1239–1253, ISSN 1364-5072.

Stroud, J.L.; Paton, G.I. & Semple, K.T. (2009). Predicting the biodegradation of target hydrocarbons in the presence of mixed contaminants in soil. *Chemosphere*, Vol. 74, No. 4, pp. 563–567, ISSN 0045-6535.

Trinidade, P.V.O.; Sobral, A.C.L.; Rizzo, S.G.F. & Leite Soriano, A.U. (2005). Bioremediation of a weathered and a recently oil-contaminated soils from Brazil: a comparasion study. *Chemosphere*, Vol. 58, No. 4, pp. 515-522, ISSN 0045-6535.

Vidali, M. (2001). Bioremediation. An overview. *Pure and Applied Cheistry*, Vol. 73, No. 7, pp. 1163–1172, ISSN 1365-3075.

Wethasinghe, C.; Yuen, S.T.S.; Kaluarachchi, J.J. & Hughes, R. (2006). Uncertainty in biokinetic parameters on bioremediation: Health risks and economic implications. *Environment International*, Vol. 32, No. 3, pp. 312 – 323, ISSN 0160-4120.

Yerushalmi, L. Rocheleau, S.; Cimpoia, R.; Sarrazin, M.; Sunahara, G.; Peisajovich, A; Leclair, G. & Guiot, S.R. (2003). Enhanced Biodegradation of Petroleum Hydrocarbons in Contaminated Soil. *Bioremediation Journal*, Vol. 7, No. 1, pp. 37–51, ISSN 1088-9868.

Chapter 4

Remediation of Contaminated Sites

Arezoo Dadrasnia, N. Shahsavari and C. U. Emenike

Additional information is available at the end of the chapter

http://dx.doi.org/10.5772/51591

1. Introduction

Oil pollution in the environment is now being taken seriously by the oil industries and as such, these companies are always looking for cost-effective methods of dealing with this pollution. The global environment is under great stress due to urbanization and industrialization as well as population pressure on the limited natural resources. The problems are compounded by drastic changes that have been taking place in the lifestyle and habits of people. The environmental problems are diverse and sometimes specific with reference to time and space. The nature and the magnitude of the problems are ever changing, bringing new challenges and creating a constant need for developing newer and more appropriate technologies.

In this context, biotechnology has tremendous potential to cater for the needs and holds hope for environmental protection, sustainability and management [1-2] While some applications such as bioremediation are direct applications of biotechnology [3-4][5], there are many which are indirectly beneficial for environmental remediation, pollution prevention and waste treatment. Large-scale pollution due to man-made chemical substances and to some extent by natural substances is of global concern now. Seepage and run-offs due to the mobile nature, and continuous cycling of volatilization and condensation of many organic chemicals such as pesticides have even led to their presence in rain, fog and snow [6].

Every year, about 1.7 to 8.8 million metric tons of oil is released into the world's water. More than 90% of this oil pollution is directly related to accidents due to human failures and activities including deliberate waste disposal [7]. PAHs are present at levels varying from 1 μg to 300 g kg^{-1} soil, depending on the sources of contamination like combustion of fossil fuels, gasification and liquefaction of coal, incineration of wastes, and wood treatment processes [8]. Incomplete combustion of organic substances gives out about 100 different polycyclic aromatic hydrocarbons (PAHs) which are the ubiquitous pollutants.

Except for a few PAHs used in medicines, dyes, plastics and pesticides, they are rarely of industrial use [9]. Some PAHs and their epoxides are highly toxic, and mutagenic even to microorganisms. About six specific PAHs are listed among the top 126 priority pollutants by the US Environmental Protection Agency. As much as the diversity in sources and chemical complexities in organic pollutants exists, there is probably more diversity in microbial members and their capabilities to synthesize or degrade organic compounds [10-11]-[12]. There are three main approaches in dealing with contaminated sites: identification of the problem, assessment of the nature and degree of the hazard, and the best choice of remedial action. The need to remediate these sites has led to the development of new technologies that emphasize the detoxification and destruction of the contaminants [13-14]-[15] rather than the conventional approach of disposal.

Remediation, whether by biological, chemical or a combination of both means, is the only option as the problem of pollution has to be solved without transferring to the future.

2. Manuscript

2.1. Measuring pollutant concentrations

The setting of soil pollution limits assumes an agreed method for measuring the concentration of a pollutant that is relevant to risk assessment across differing soil types. Limits are generally expressed in terms of 'total' concentrations as there is no consensus on alternative [16] methods more directly related to biological or environmental risk. Yet, assessing the bioavailability of soil pollutants is an essential part of the process of risk assessment and of determining the most appropriate approach to remediation [17]. With developing non exhaustive solvent extraction procedures that consistently predict the bioavailability of organic contaminants across a range of soil conditions [17].

As an alternative to extraction, solid-phase micro-extraction uses adsorbents added to soil–water slurries aiming to mimic the accessibility of organic contaminants to microorganisms. In relation to the assessment of risks to human health, much work is currently underway to develop physiologically based extraction tests; however, progress made in this respect for inorganic pollutants has not been matched by that for organic pollutants [18]. In recent years, there has been a growth in the use of onsite assays to improve decision making regarding the extent of pollution in batches of potentially polluted materials and, therefore, the need for treatment or disposal. In many cases, these new measurements are based on enzyme-linked immunosorbent assays linked to spectroscopy.

Specific assays have, for example, been developed for pentachlorophenol [19] and PAHs [20].Whilst these methodologies can provide useful supplementary and 'real-time' information on pollutant concentration variability in the field, care must be taken when extrapolating findings from the very small samples used in these assays to bulk soil properties.

Various microbiological assays have been proposed as indicators of pollutant bioavailability. Biosensors have been widely deployed to provide fast, cost-effective monitoring of pollutants and their biological toxicity.

2.2. Environmental pollution and biological treatments

The problems of environment can be classified into the following subheads as most of the problems can be traced to one or more of the following either directly or indirectly: Waste generation (sewage, wastewater, kitchen waste, industrial waste, effluents, agricultural waste, food waste) and use of chemicals for various purposes in the form of insecticides, pesticides, chemical fertilizers, toxic products and by-products from chemical industries (Fig 1). Waste generation is a side effect of consumption and production activities and tends to increase with economic advance. What is of concern is the increased presence of toxic chemicals such as halogen aliphatics, aromatics, polychlorinated biphenyls and other organic and inorganic pollutants which may reach air, water or soil and affect the environment in several ways, ultimately threatening the self-regulating capacity of the biosphere [5]-[21]-[22].

They may be present in high levels at the points of discharge or may remain low but can be highly toxic for the receiving bodies. The underground water sources are increasingly becoming contaminated. For example, the underground water sources have been permanently abandoned in the valley of the River Po in north Italy due to industrial pollution. Some substances may reach environment in small concentrations but may be subjected to biomagnification or bioaccumulation up the food chain, wherein their concentrations increase as they pass through the food chain [23]-[24]-[25-26].

All the more, rapid developments in understanding activated sludge processes and wastewater remediation warrant exploitation of different strategies for studying their degradation and some of the biological remediation terminologies such as bioleaching, biosorption, bioaugmentation, biostimulation, biopulping, biodeterioration, biobleaching, bioaccumulation, biotransformation and bioattenuation are being actively researched on [27]. Enzyme technology has equally been receiving increased attention. Hussain et al. (2009) have reviewed the biotechnological approaches for enhancing the capability of microorganisms and plants through the characterization and transfer of pesticide-degrading genes, induction of catabolic pathways, and display of cell surface enzymes[28], while Theron et al. (2008) have performed a thorough review of nanotechnology, the engineering and art of manipulating matter at the nanoscale (1–100 nm), and have highlighted the potential of novel nanomaterials for treatment of surface water, groundwater, and wastewater contaminated by toxic metal ions, organic and inorganic solutes, and microorganisms [29]. Husain et al. (2009) have analyzed the role of peroxidases in the remediation and treatment of a wide spectrum of aromatic pollutants[28].

Remediation approaches encompass applied physical, chemical and biological environmental sciences. The aim of this chapter will be to illustrate current understanding of the scientific principles underlying soil remediation and some of the challenges to their successful application. Remediation approaches that isolate treated soils are site rather than soil remediation technologies. These approaches, and the treatments that result in the destruction of soil function, will be referred to only in passing.

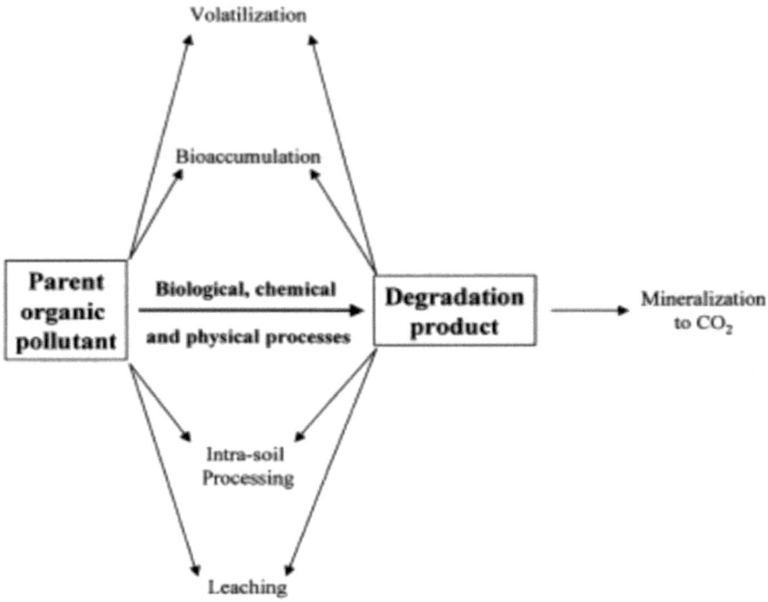

Figure 1. Summary of environmental fates on organic pollutants in soil.

2.2.1. Bioremediation

Interest in the microbial biodegradation of pollutants has intensified in recent years as humanity strives to find sustainable ways to clean up contaminated environments. Bioremediation, which is the use of microorganisms consortia or microbial processes to degrade and detoxify environmental contaminants [30]. It is also amongst these new technologies which derives its scientific justification from the emerging concept of Green Chemistry and Green Engineering, and is a fast growing promising remediation technique increasingly being studied and applied in practical use for pollutant clean-up.

Bioremediation techniques have been used for decontamination of surface and subsurface soils, freshwater and marine systems, soils, groundwater and contaminated land ecosystems. However, the majority of bioremediation technologies initially developed were to treat petroleum hydrocarbon contamination to immobilize contaminants or to transform them to chemical products no longer hazardous to human health and the environment. Where contaminants pose no significant risk to water supply or surface water bodies, biodegradation products will include carbon dioxide, water and other compounds with little deleterious effects on the environment [31].

Bioremediation of soils or any site may be enhanced by fertilizing (adding nutrients such as carbon, nitrogen and phosphorous) and/or seeding with suitable microbial populations. These days, using organic wastes is bioremediation process is going to be new method as a option of enhancing and motivating of microorganism to break down of organic compounds [32-33]. This is enhanced or engineered bioremediation. Intrinsic bioremediation, which utilizes existing microbial communities, is often the most cost effective method available for land decontamination. Even in the most contaminated soils, indigenous microbial activity can be enough to clean the soil effectively. Bioremediation techniques are cost effective as compared to other technologies. Biological treatments compare favorably with alternative methods. Treatment periods generally last from 2 to 48 months, about the same for chemical or thermal methods. Physical processes (soil washing and soil vapour extraction) are faster, rarely lasting more than 1 year. Solidification is almost instantaneous.

Bioremediation (when used in solution) does not require environmentally damaging processes such as chemicals or heat treatment. It has beneficial effects upon soil structure and fertility, but with limitation on its effectiveness. These limitations may be summarized as follows:

- Space requirements
- Monitoring difficulties
- Extended treatment time

2.2.1.1. Bioremediation technologies

Bioremediation technologies can be broadly classified as ex situ or in situ. Table 1 summarizes the most commonly used bioremediation technologies. Ex situ technologies are those treatment modalities which involve the physical removal of the contaminated to another area (possibly within the site) for treatment.

Bioreactors, land farming, anaerobic digestion, composting, biosorption and some forms of solid-phase treatment are all examples of ex situ treatment techniques. In contrast, in situ techniques involve treatment of the contaminated material in place. Bioventing for the treatment of the contaminated soil and biostimulation of indigenous aquifer microorganisms are examples of these treatment techniques. Although some sites may be more easily controlled and maintained with ex situ configurations [34].

Bioaugmentation	Addition of bacterial cultures to a contaminated medium frequently used in bioreactors and ex situ systems
Biofilters	Use of microbial stripping columns to treat air emission
Biostimulation	Stimulation of indigenous microbial populations in soils and/or ground water
Bioreactors	Biodegradation in a container or reactor
Bioventing	Method of treating contaminated soils by drawing oxygen through the soil to stimulate microbial growth and activity

Table 1. Bioremediation treatment technologies

For example, many sites are located in industrial/ commercial areas, and these sites normally consist of numerous structures interconnected by concrete and asphalt. These physical barriers would make excavation extremely difficult, and if the contamination is deep in the subsurface, excavation becomes too expensive. As a result of these physical barriers, the required excavation efforts may make ex situ biotreatment impracticable. Other factors could also have an impact on the type of treatment. At a typical site, the contamination is basically trapped below the surface.

To expose the contamination to the open environment through excavation can result in potential health and safety risks [34]. In addition, the public's perception of the excavation of contaminants could be negative, depending on the situation. All of these conditions clearly favor in situ biotreatment. Nonetheless, the key is to carefully consider the parameters involved with each site before evaluating which technique to use [34].

2.2.2. Land farming

This technology involves the application of contaminated material that has been excavated onto the soil surface and periodically tilled to mix and aerate the material [35-36]. The contaminants are degraded, transformed and immobilized by means of biotic and abiotic reactions. Sometimes, in cases of very shallow contamination, the top layer of the site may simply be tilled without requiring any excavation. Liners or other methods may be used to control leachate. This technology is designed primarily to treat soil contamination by fuels, PAHs, non-halogenated VOCs, SVOCs, pesticides, and herbicides. The process may be applied to halogenated organics, but is less effective.

simple and inexpensive, it does require large space, and reduction in contaminant concentrations may sometimes be due to volatilization rather than biodegradation [37-38]. Marı´n et al. (2005) assessed the ability land farming to reduce the total hydrocarbon content added to soil with refinery sludge in low rain and high temperature conditions [39]. It was seen that 80% of the hydrocarbons were eliminated in 11 months, half of this reduction taking place during the first 3 months.

2.2.3. Phytoremediation

Using plants in soil and groundwater remediation (i.e., phytoremediation) is a relatively new concept and the technology has yet to be extensively proven in the marketplace. However, the potential of phytoremediation for cheap, simple and effective soil and groundwater remediation is generating considerable interest.

Phytoremediation may be used for remediation of soil and groundwater contaminated with toxic heavy metals, radio nuclides, and organic contaminants such as chlorinated solvents, BTEX compounds, non-aromatic petroleum hydrocarbons, nitro toluene ammunition wastes, and excess nutrients [40]. Other applications of phytoremediation include Land fill caps, buffer zones for agricultural runoff and even drinking water and industrial wastewater treatment. Phytoremediation may also be used as a final polishing step, in conjunction with other treatment technologies. While indeed promising, the applicability of phytoremediation is limited by several factors. First, it is essential that the contaminated site of interest is able to support plant growth. This requires suitable climate, soil characteristics such as pH and texture, and adequate water and nutrients. Second, because plant roots only go so deep, phytoremediation is practical only in situations where contamination is shallow (less than 5 m), although in some situations with deeper contamination it may be used in conjunction with other technologies. Third, since the time requirements for phytoremediation are sometimes long relative to some conventional technologies such as land filling and incineration, it is not suitable for situations requiring rapid treatment. Plants facilitate remediation via several mechanisms (Fig 2):

1. Direct uptake, and incorporation of contaminants into plant biomass
2. Immobilization, or Phytostabilization of contaminants in the subsurface
3. Release plant enzymes into the rhizosphere that act directly on the contaminants
4. Stimulation of microbial mediated degradation in the rhizosphere

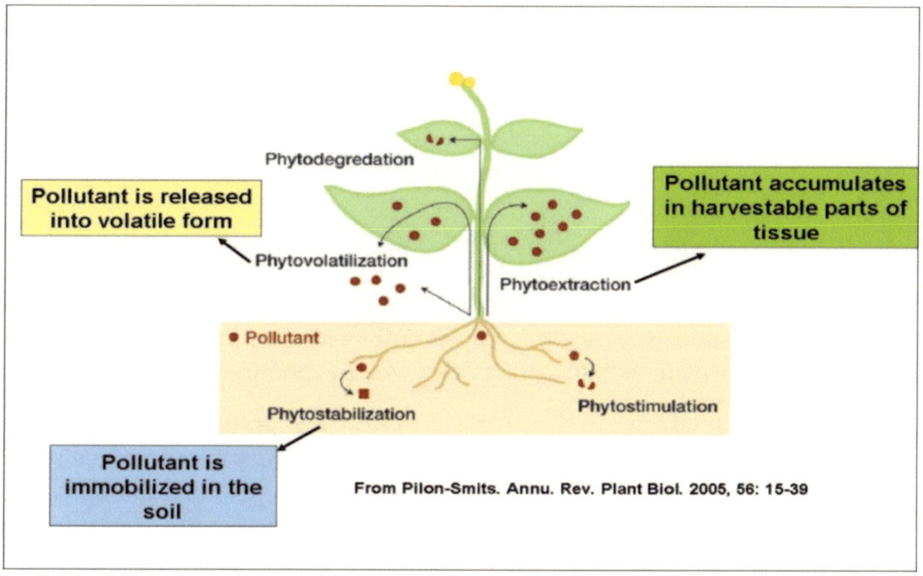

Figure 2. Phytoremediation mechanisms.

2.2.4. *Biopiling*

Biopiles piles are a form of soil treatment where bulking agents, nutrients, and water are added. However, static piles are not mixed and temperatures are usually near ambient. Aeration can be passive or forced by applying a vacuum or blowing air through the pile. Bulking agents used are usually made up of manure or compost, which supports a larger microbial population than soil and provides inorganic nutrients, and relatively inert materials such as sawdust, wood chips, or compost. Water is added periodically, as needed to sustain the microbial population [41-42].

2.2.5. *Composting*

Composting is an aerobic process that relies on the actions of microorganisms to degrade organic materials, resulting in the thermo genesis and production of organic and inorganic compounds. The metabolically generated heat is trapped within the compost matrix, which leads to elevations in temperature, a characteristic of composting. In deed composting is the biochemical degradation of organic materials to a sanitary, nuisance-free, humus-like material [43]. Composting has been defined as a controlled microbial aerobic decomposition process with the formation of stabilized organic materials that may be used as soil conditioner [44]. The main factors in the control of a composting process include environmental parameters (temperature, moisture content, pH and aeration) and substrate nature parameters (C/N ratio, particle size, and nutrient content) [45-46].

Various factors correlate with each other physically, chemically and biologically in complicated composting processes. A slight change in a single factor may cause a drastic avalanche of metabolic and physical changes in the overall process. In other words, there may be extremely strong non-linearities involved in these processes [47]. These processes occur in matrix of organic particles and interconnected pores, and the pores are partially filled with air, aqueous solution, or a combination of the two. A multitude of microorganisms and their enzymes is responsible for the biodegradation process [48], resulting in a complex biochemical–microbial system.

2.2.6. Electrokinetic remediation

Electrokinetic treatment is emerging and innovative technology to complement traditional technology limitations and to treat fine-grained soils. Electrokinetic technology evaluated most suitable to remove contaminants effectively from low permeability clayey soil.

In situ electrokinetic remediation can be applied to treat low permeable soils contaminated with heavy metals, radionuclides and selected organic pollutants. The principle behind this method is the application of a low level direct current electric potential through electrodes, which are placed into the contaminated soil. Ionic contaminants are transported to the oppositely charged electrode by electromigration. Additionally, electroosmotic flow provides a driving force for the movement of soluble contaminants [49].

Although the technology has been known and utilised for more than a decade, application to removal of hydrophobic and strongly adsorbed pollutants such as PAHs especially from low permeability soils is recent. Solubilising agents are therefore used in these cases to enhance the removal efficiency of PAHs [50].

2.2.7. Photocatalytic degradation

The photocatalytic degradation process uses photocatalysts to promote oxidising reactions which destroy organic contaminants in the presence of light radiation. The technology has been widely established for treatment of wastewater, and recently, its application has extended to treatment of contaminated soils.

Zhang et al. [51]conducted a comprehensive study of the photocatalytic degradation of phenanthrene, pyrene and benzo(a)pyrene on soil surfaces using titanium dioxide (TiO_2) under UV light. Compared to the absence of catalyst, the addition of TiO_2 as catalyst revealed that TiO_2 accelerated the photodegradation process of all three PAHs, with benzo(a)pyrene being degraded the fastest. Nonetheless, variation in TiO_2 concentration from 0.5 to 3wt. % did not provide any significant effect on PAH degradation. Under distinct UV wavelengths, photocatalytic degradation rates of PAHs were different. Soil pH was discovered to affect the process whereby the highest pyrene and benzo(a)pyrene degradation rates were obtained at acidic conditions, while phenanthrene was most significantly degraded at alkaline conditions. Additionally, the presence of humic acid in soil was found to enhance PAH photocatalytic degradation by sensitising radicals capable of oxidizing PAHs.

Rababah and Matsuzawa [52]developed a recirculating-type photocatalytic reactor assisted by the oxidising agent H_2O_2 solution to treat soil spiked with fluranthene. It was observed that the degradation efficiency of fluoranthene was 99% in the presence of both TiO_2 and H_2O_2 compared to a lower degradation efficiency of 83% in the presence of TiO_2 alone.

2.3. Physico- chemical treatments

2.3.1. Solidification and Stabilization

Solidification/Stabilization (S/S) is one of the top five source control treatment technologies used at Superfund remedial sites, having been used at more than 160 sites. "Solidification" refers to a process in which materials are added to the waste to produce an immobile mass. This may or may not involve a chemical bonding between the toxic contaminant and the additive. "Stabilization" refers to converting a waste to a more chemically stable form. This conversion may include solidification, but it almost always includes use of physicochemical reaction to transform the contaminants to a less toxic form [53-54].

Solidification is a technique that encapsulates hazardous waste into a solid material of high structural integrity. Solidifying fine waste particles is termed microencapsulation; macro encapsulation solidifies wastes in large blocks or containers. Stabilization technologies reduce a hazardous wastes solubility, mobility, or toxicity. Solidification and stabilization are effective for treating soils containing metals, asbestos, radioactive materials, in organics, corrosive and cyanide compounds, and semi-volatile organics. Solidification eliminates free liquids, reduces hazardous constituent mobility by lowering waste permeability, minimizes constituent leach ability, and provides stability for handling, transport, and disposal [55].

2.3.2. Soil vapour extraction

In cases where the contaminants are volatile, a venting and ex-situ gas treatment system can be applied. Soil vapour extraction is a technology that has been proven effective in reducing concentrations of VOC and certain semi-volatile organic compounds. Principally, a vacuum is applied to the soil matrix to create a negative pressure gradient that causes movement of vapors toward extraction wells. Volatile contaminants are readily removed from the subsurface through the extraction wells. The collected vapors are then treated and discharged to the atmosphere or where permitted, reinjected to the subsurface [56-57].

2.3.3. *Soil washing*

Soil washing is an *ex situ* treatment technology for the remediation of contaminated soil. It has been applied to a variety of inorganically, organically, and even radioactively contaminated soils. Although it is a well established technology in continental Europe and North America, there are very few applications in the UK.

The selection of soil washing for a particular contamination problem will depend on a variety of factors. Particularly important is whether the contamination is specific to particular groups of particles within the soil and whether these particles can be removed from the contaminant-free bulk of particles by physical or physico-chemical processes [58].

Contamination can occur on or in soil particles in a variety of ways. Six types of association are identified:

- *Adsorbed contamination*. Contaminants may be adsorbed to particles and, in many cases, this adsorption may be preferential to particular particle types. For example, the adsorption of inorganic or organic contaminants on peaty organic fraction or on clay particles.
- *Discrete particles*. Some contaminants may occur within the soil matrix as discrete particles that are not necessarily associated with soil particles. Contaminants of this type can include discrete metal grains or metal oxides, tar balls and some waste materials (e.g. used catalyst fragments).
- *Coatings*. Contaminants may occur as coatings on individual particles that have resulted from precipitation of the contaminant from solution. For example, metal salts and iron oxides can precipitate on sand particles.
- *Liquid or semi-liquid coating*. Liquid or semi-liquid viscous substances may occur as coatings around individual soil particles. Contaminants of this type can include oils, tars and some other organic contaminants.
- *Liquid or semi-liquid coating*. Liquid or semi-liquid viscous substances may occur as coatings around individual soil particles. Contaminants of this type can include oils, tars and some other organic contaminants.
- *Internal contamination within pores*. Contamination may also occur inside individual grains but within the pore structure. Here it may be adsorbed (e.g. heavy metal or organic contamination), occur as a coating to the pore walls (e.g. inorganic compound precipitated from solution) or occur within and possibly occlude the pores (e.g. contaminants such as mineral oils).
- *Part of individual grains*. Contamination may occur within the matrix of an individual grain, or as part of an individual grain. Heavy metal contamination may occur in this way, for example, in slags where the heavy metal can occur within the vitrified matrix or associated with specific mineral phases such as magnetite.

Soil washing technology involves mixing the solvent (water) and contaminated soil in an extractor vessel [41, 59]. The mixing dissolves the organic contaminant into the solvent. Solvent and dissolved contaminants are then placed in a separator where the solute and solvent are separated and treated. The soils can be stockpiled, tested and used as inert material (Fig 3).

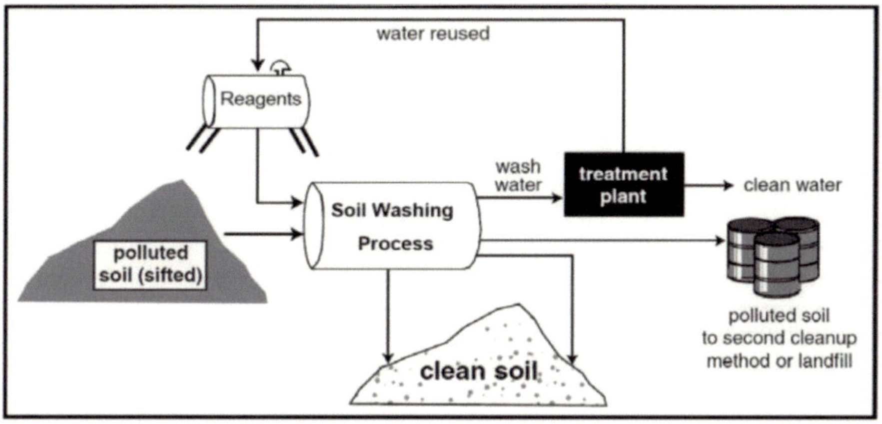

Figure 3. Schematic diagram of soil washing.

2.3.4. *Air-sparging*

Air sparging is an in situ technology in which air is injected through a contaminated aquifer. Air-sparing stimulates aerobic biodegradation of contaminated groundwater by delivery of oxygen to the subsurface [60]-[61]. This is accomplished by injecting air below the water table. This technology is designed primarily to treat groundwater contamination by fuels, non-halogenated VOCs, SVOCs, pesticides, organics, and herbicides.

Air sparing has also been demonstrated to be an innovative groundwater remediation technology capable of restoring aquifers that have been polluted by volatile and (or) biodegradable contaminants, such as petroleum hydrocarbons. The process may be applied to halogenated organics, but is less effective (Fig 4).

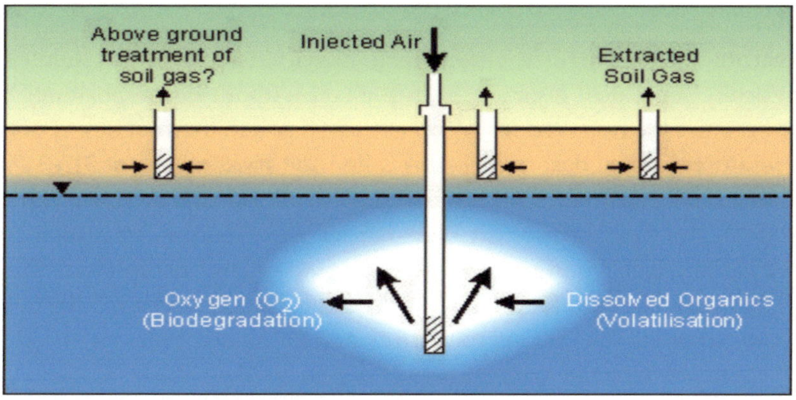

Figure 4. A schematic diagram illustrating method of air sparging.

Air-sparing can cost less than $1 per 1,000 l in favorable situations and tends to be among the cheapest remedial alternatives when applicable. The technology uses simple, inexpensive, low-maintenance equipment that can be left unattended for long periods of time. Also, the technology tends to enjoy good public acceptance. The technology requires the presence of indigenous organisms capable of degrading the contaminants of interest, as well as nutrients necessary for growth. Also, it is necessary that the contaminants be available to the organisms, and not tightly sorbed to soil particles. Air sparing is not applicable in sites where high concentrations of inorganic salts, heavy metals, or organic compounds are present, as hinder microbial growth.

Excavation (and removal) is a fundamental remediation method involving the removal of contaminated soil/media, which can be shipped off-site for treatment and/or disposal, or treated on-site when contaminants are amenable to reliable remediation techniques.

Excavation is generally utilized for localized contamination and point source and is also used for the removal of underground structures that are out of compliance or have been identified as a potential or actual point source of contamination. The limiting factor for the use of excavation is often represented by the high unit cost for transportation and final offsite disposal. EPA (1991) further stated some limiting factors that may limit the applicability and effectiveness of the process to include:

i. Generation of fugitive emissions may be a problem during operations.
ii. The distance from the contaminated site to the nearest disposal facility will affect cost.
iii. Depth and composition of the media requiring excavation must be considered.
iv. Transportation of the soil through populated areas may affect community acceptability.

In this respect, the on-site removal and treatment can often yield significant savings and, in addition, the treated soil may have beneficial secondary use (e.g. as construction fill or road base material) at the same site.

2.4. Thermal treatment

2.4.1. Thermal desorption

Thermal desorption technology is based on a physical separation system. The process desorbs (physically separates) organics from the soil without decomposition. Volatile and semi-volatile organics are removed from contaminated soil in thermal desorbers at 95-315°C for low-temperature thermal desorption (also called soil roasting), or at 315-340°C for high-temperature thermal desorption. To transport the volatilized organics and water to the gas treatment system, the process uses an inert carrier gas. The gas treatment units can be condensers or carbon adsorption units, which will trap organic compounds for subsequent treatment or disposal. The units can also be afterburners or catalytic oxidizers that destroy the organic constituents. The bed temperatures and residence times of the desorbers are designed to volatilize selected contaminants, not to oxidize them. Certain less volatile compounds may not be volatilized at low temperatures [62-63].

2.4.2. Incineration

For the remediation of soils polluted with organic compounds, incineration is the most widely used method. This method is very expensive and generates problems with air emissions and noise [64]. Incineration technology is intended to permanently destroy organic contaminants. Incineration is a complex system of interacting pieces of equipment and is not just a simple furnace. It is an integrated system of components for waste preparation, feeding, combustion, and emissions control. Central to the system is the combustion chamber, or the incinerator. There are four major types of incinerator: rotary kiln, fluidized bed, liquid injection, and infrared.

2.5. Novel remediation techniques

2.5.1. Nanotechnology and remediation

Nanotechnology has contributed to the development of a great diversity of materials as those used in electronic, optoelectronic, biomedical, pharmaceutical, cosmetic, energy, catalytic, and materials applications. As a general definition, nanotechnology is involved with objects on the nano scale, or materials measuring between 1 and 100 nm [65]. In future, modification and adaptation of nanotechnology will extend the quality and length of life [66]. The social benefits are significant from nanomaterials and the new products are applicable to information technology, medicine, energy, and environment. The emergence of nanotechnology presents a number of potential environmental benefits.

2.5.2. Steam stripping

The steam stripping method is based on a mass transfer concept, which is used to move volatile contaminants from water to air. Steam is injected through an injection well into the soil to vaporize volatile and semi-volatile contaminants [67]. The contaminated vapour steam is removed by vacuum extraction, and the contaminants are then captured through condensation and phased separation processes [68].

2.5.3. Dehalogenation

Dehalogenation of organic compounds is chemical displacement of a chlorine molecule and resulting reduction of toxicity.

2.5.4. Chemical reduction/oxidation

Chemical reduction/oxidation is a chemical conversion of hazardous contaminants to non-hazardous or less toxic compounds. The result is a more stable, less mobile and/or inert material [69].

2.5.5. Ultraviolet (UV) oxidation

Ultraviolet (UV) oxidation technology uses UV radiation, ozone, or hydrogen peroxide to destroy or detoxify organic contaminants as water flows into a treatment tank. The reaction products are dechlorinated materials and chlorine gas [70-71].

2.5.6. *Supercritical fluids extraction*

Supercritical fluids are materials at elevated temperature and pressure that have properties between those of a gas and a liquid. Under these conditions, the organic contaminant readily dissolves in the supercritical fluid. Supercritical fluids processes represent emerging technologies in the site remediation field. Few full-scale applications of Supercritical fluids are currently in existence [72-73].

3. Conclusion

A number of organic pollutants, such as PAHs, PCBs and pesticides, and inorganic pollutants are resistant to degradation and represent an ongoing toxicological threat to both wildlife and human beings. Bioremediation has grown into a green, attractive and promising alternative to traditional physico-chemical techniques for the remediation of hydrocarbons at a contaminated site, as it can be more cost-effective and it can selectively degrade the pollutants without damaging the site or its indigenous flora and fauna. However, bioremediation technologies have had limited applications due to the constraints imposed by substrate and environmental variability, and the limited biodegradative potential and viability of naturally occurring microorganisms. For the development of remediation processes to succeed commercially, it is essential to link different disciplines such as microbial ecology, biochemistry and microbial physiology, together with biochemical and bioprocess engineering.

In short, the key to successful remediation resides in continuing to develop the scientific and engineering work that provides the real bases for both the technology and its evaluation; and simultaneously in explaining and justifying the valid reasons which allow scientists and engineeres to actually use these technologies for the welfare and safety of a public which is more and more concerned about the environment and its protection.

Author details

Arezoo Dadrasnia and C. U. Emenike
Institute of Biological Science, University of Malaya, Kuala Lumpur, Malaysia

N. Shahsavari
School of Biological Science, University Sains Malaysia, Penang, Malaysia
Hajiabad branch, Islamic Azad University, Hajiabad, Hormozgan, Iran

Acknowledgement

We wish to express our deepest gratitude to all the researchers whose valuable data as reported in their respective publications and cited in this chapter have been of considerable significance in adding substance. We are also grateful to our other colleagues and the anonymous reviewers whose constructive criticisms have benefited the manuscript, and brought it to its present form.

4. References

[1] Hatti-Kaul R, To¨rnvall U, Gustafsson L, B.r. P, Industrial biotechnology for the production of bio-based chemicals—a cradle-to-grave perspective., Trends Biotechnol, 25 (2007) 119–124.
[2] Azadi H, H. P, Genetically modified and organic crops in developing countries: a review of options for food security, Biotechnol Adv 28 (2010) 160–168.
[3] S. Koenigsberg, T. Hazen, A. Peacock, Environmental biotechnology: a bioremediation perspective, Remed J 15 (2005) 5–25.
[4] Dowling DN, D. SL, Improving phytoremediation through biotechnology, Curr Opin Biotechnol 20 (2009) 204–209.
[5] Sen R, Chakrabarti S, Biotechnology-applications to environmental remediation in resource exploitation., Curr Science Press, Beijing,, 97 (2009) 768–775.
[6] Dubus IG, Hollis JM, B. CD, Pesticides in rainfall in Europe, Environ Pollut 110 (2000) 331–344.
[7] Zhu X, Venosa AD, Suidan MT, L. K, Guidelines for the bioremediation of marine shorelines and freshwater wetlands. Cincinnati, OH: US Environmental Protection Agency;, http://www.epa.gov/oilspill/pdfs/bioremed.pdf., (2001).
[8] Bamforth SM, S. I., Bioremediation of polycyclic aromatic hydrocarbons: current knowledge and future directions., J Chem Technol Biotechnol 80 (2005) 723–736.
[9] U. EPA., . Great Lakes National Program Office. Realizing remediation: a summary of contaminated sediment remediation activities in the Great Lakes Basin. US Environmental Protection Agency, http://www.epa.gov/greatlakes/sediment/realizing/. (1998).
[10] Ramakrishnan B, Megharaj M, Venkateswarlu K, S.N. Naidu R, The impacts of environmental pollutants on microalgae and cyanobacteria, Crit Rev Environ Sci Technol, (2010) 699–821.
[11] Ramakrishnan B, Megharaj M, Venkateswarlu K, Sethunathan N, N. R., Mixtures of environmental pollutants: effects on microorganisms and their activities, Rev Environ Contam Toxicol, 211 (2011) 63–120.
[12] W. K., Microorganisms relevant to bioremediation, Curr Opin Biotechnol, 12 (2001) 237–241.
[13] Wang G-D, C. X-Y, Detoxification of soil phenolic pollutants by plant secretory enzyme, phytoremedation, Humana Press, Totowa, , (2007) 49–57.
[14] W. R., Relevance of PCDD/PCDF formation for the evaluation of POPs destruction technologies—review on current status and assessment gaps. , Chemosphere 67 (2007) 109–117.
[15] Kulkarni PS, Crespo JG, A. CAM, Dioxins sources and current remediation technologies—a review. , Environ Int. , 34 (2008) 139–153.
[16] McLaughlin MJ, Zarcinus BA, Stevens DP, Cook N, Soil testing for heavy metals, Commun Soil Sci Plant Anal, 31 (2000) 1661–1700.
[17] Semple KT, Morriss AWJ, Paton GI, Bioavailability of hydrophobic organic contaminants in soils: fundamental concepts and techniques for analysis. , Eur J Soil Sci 54 (2003) 809–818.
[18] Ruby MV, Bioavailability of soil-borne chemicals: abiotic assessment tools. , Human Ecol Risk Assess 10 (2004) 647–656.

[19] Li M, Tsai SF, Rosen SM, Wu RS, Reddy KB, DiCesare J, Salamone SJ, Preparation of pentachlorophenol derivatives and development of a microparticle-based on-site immunoassay for the detection of PCP in soil samples, J Agric Food Chem 49 (2001) 1287–1292.

[20] S.M. Knopp D, Vaananen V, Niessner R Determination of polycyclic aromatic hydrocarbons in contaminated water and soil samples by immunological and chromatographic methods., Environ Sci Technol 34 (2000) 2035–2041.

[21] Prasad MNV, Freitas H, Fraenzle S, Wuenschmann S, M. B, Knowledge explosion in phytotechnologies for environmental solutions., Environ Pollut, 158 (2010) 18–23.

[22] Beltrame MO, De Marco SG, M. JE., Effects of zinc on molting and body weight of the estuarine crab Neohelice granulata (Brachyura: Varunidae). , Sci Total Environ Exp Bot, 408 (2010) 531–536.

[23] Davies OA, Allison ME, U. HS, Bioaccumulation of heavy metals in water, sediment and periwinkle (Tympanotonus fuscatus var radula) from the Elechi Creek, Niger Delta., Afr J Biotechnol 5(2006) 968–973.

[24] Kelly BC, Ikonomou MG, Blair JD, Morin AE, G. FAPC, Food web-specific biomagnification of persistent organic pollutants, Science Press, Beijing,, 317 (2007) 236–239.

[25] Fatemi MH, Baher E, A novel quantitative structureactivity relationship model for prediction of biomagnification factor of some organochlorine pollutants, Mol Divers 13 (2009) 343–352.

[26] Takeuchi I, Miyoshi N, Mizukawa K, Takada H, Ikemoto T, Omori K, Tsuchiya K, Biomagnification profiles of polycyclic aromatic hydrocarbons, alkylphenols and polychlorinated biphenyls in Tokyo Bay elucidated by d13C and d15N isotope ratios as guides to rophic web structure. , Mar Pollut Bull, 58 (2009) 663–671.

[27] Whiteley CG, Lee D-J, Enzyme technology and biological remediation, Enzym Microb Technol 38 (2006) 291–316.

[28] Husain Q, Husain M, Kulshrestha Y, Remediation and treatment of organopollutants mediated by peroxidases: a review, Crit Rev Biotechnol 29 (2009) 94–119.

[29] Theron J, Walker JA, Cloete TE, Nanotechnology and water treatment: applications and emerging opportunities., Crit Rev Microbiol 34 (2008) 43–69.

[30] B. Zhao, C.L. Poh, Insights into environmental bioremediation by microorganisms through functional genomics and proteomics, PROTEOMICS, 8 (2008) 874-881.

[31] Baker DB, Conradi MS, N. RE, Explanation of the high-temperature relaxation anomaly in a metal-hydrogen system., Phys Rev B 49 (1994) 11773–11782.

[32] A. Dadrasnia, P. Agamuthu, Enhanced Degradation of Diesel-Contaminated Soil using Organic Wastes, Malaysian Journalof Science, 29 (2010) 225-230.

[33] P. Agamuthu, O.P. Abioye, A.A. Aziz, Phytoremediation of soil contaminated with used lubricating oil using Jatropha curcas, Journal of Hazardous Materials, 179 (2010) 891-894.

[34] Talley WF, S. PM, Roadblocks to the implementation of biotreatment strategies., Ann NY Acad Sci, (2006) 16–29.

[35] Maciel BM, Santos ACF, Dias JCT, Vidal RO, Dias RJC, Gross E, Cascardo JCM, R. RP, Simple DNA extraction protocol for a 16S rDNA study of bacterial diversity in tropical landfarm soil used for bioremediation of oil waste, Genet Mol Res 8(2009) 375–388.

[36] Harmsen J, Rulkens WH, Sims RC, Rijtema PE, Z. AJ, Theory and application of landfarming to remediate polycyclic aromatic hydrocarbons and mineral oilcontaminated sediments; beneficial reuse. , J Environ Qual, 36 (2007) 1112–1122.

[37] Sanscartier D, Reimer K, Zeeb B, G. K, Management of hydrocarbon—contaminated soil through bioremediation and landfill disposal at a remote location in Northern Canada, Can J Civil Eng 37 (2010) 147–155.

[38] Souza TS, Hencklein FA, Angelis DF, Gonc¸alves RA, Fontanetti, CS, The Allium cepa bioassay to evaluate landfarming soil, before and after the addition of rice hulls to accelerate organic pollutants biodegradation, Ecotoxicol Environ Saf 72 (2009) 1363–1368.

[39] Marı´n JA, Hernandez T, G. C, Bioremediation of oil refinery sludge by landfarming in semiarid conditions: influence on soil microbial activity., Environ Res 98 (2005) 185–195.

[40] Schnoor JL, Licht LA, Mc Cutcheon SC, Wolf NL, C. LH, Phytoremediation of organic and nutrient contaminants., Environ Sci Technol 29 (1995) 317–323.

[41] Lucian Vasile Pavel, M. Gavrilescu, overview of ex situ decontamination techniques for soil cleanup, Environmental Engineering and Management, 7 (2008) 815-834.

[42] R. Iturbe, C. Flores, C. Chavez, G. Bautista, L. Torres, Remediation of contaminated soil using soil washing and biopile methodologies at a field level, Journal of Soils and Sediments, 4 (2004) 115-122.

[43] Kulcu R, Y. O, Determination of aeration rate and kinetics of composting some agricultural wastes, Bioresour Technol 93 (2004) 49–57.

[44] Negro MJ, Solano PC, C. J, Composting of sweet sorghum bagasse with other wastes, Bioresour Technol, 67 (1999) 89–92.

[45] Diaz MJ, Madejon E, Lopez F, Lopez R, C. F, Optimization of the rate vinasse/grape marc for co-composting process, Process Biochem 37 (2002) 1143–1150.

[46] Artola A, Barrena R, Font X, Gabriel D, Gea T, Mudhoo A, S.n. A, Composting from a sustainable point of view: respirometric indices as a key parameter. In: Martı ´n-Gil J (ed) Compost II,, dynamic soil dynamic plant, 3 (2009) 1-16.

[47] Seki H, Stochastic modeling of composting process with batch operation by the Fokker–Planck equation, Trans ASAE 43 (2000) 169–179.

[48] Fogarty AM, T. OH, Microbiological degradation of pesticides in yard waste composting., Microbiol Rev Am Soc Microbiol 55 (1991) 225–233.

[49] H.I. Gomes, C. Dias-Ferreira, A.B. Ribeiro, Electrokinetic remediation of organochlorines in soil: Enhancement techniques and integration with other remediation technologies, Chemosphere, 87 (2012) 1077-1090.

[50] D. Huang, Q. Xu, J. Cheng, X. Lu, H. Zhang, Electrokinetic Remediation and Its Combined Technologies for Removal of Organic Pollutants from Contaminated Soils, Int. J. Electrochem. Sci., 7 (2012) 4528 - 4544.

[51] L. Zhang, P. Li, Z. Gong, X. Li, Photocatalytic degradation of polycyclic aromatic hydrocarbons on soil surfaces using TiO2 under UV light,, J. Hazard. Mater, 158 (2008) 478–484.

[52] A. Rababah, S. Matsuzawa, Treatment system for solid matrix contaminated with fluoranthene. II—Recirculating photodegradation technique,, Chemosphere, 46 (2002) 49-57.

[53] U.S.E.P. Agency, Solidification/Stabilization Resource Guide, U.S. Environmental Protection Agency, (1999) 1-91.

[54] S. Pensaert, Immobilisation, Stabilisation, Solidification a New Approach for the Treatment of Contaminated Soils.Case studies: London Olympics & Total Ertvelde, 15de Innovatieforum Geotechniek, (2008) 1-14.

[55] U.S.E.P. Agency, Technology Performance Review: Selecting and Using Solidification/Stabilization Treatment for Site National Service Center for Environmental Publications (NSCEP), (2009) 1-28.
[56] S.S. Suthersan, Soil Vapor Extraction, CRC Press LLC, (1999) 1-65.
[57] A. Soares, J. Albergaria, V. Domingues, C. Alvim-Ferraz Mda, C. Delerue-Matos, Remediation of soils combining soil vapor extraction and bioremediation: benzene., Chemosphere, 80 (2010):823-828.
[58] R.A. Griffiths, Soil-washing technology and practice, Journal of Hazardous Materials, 40 (1995) 175–189.
[59] M.T. Balba, N. Al-Awadhi, R. Al-Daher, Bioremediation of oil-contaminated soil: microbiological methods for feasibility assessment and field evaluation, Journal of Microbiological Methods, 32 (1998) 155-164.
[60] Johnson RL, Johnson PC, McWhorter DB, Hinchee RE, G. I, An overview of in situ air sparging, Ground Water Monit Remed 13 (2007) 127-135.
[61] T. Y-J, Air distribution and size changes in the remediated zone after air sparging for soil particle movement., J Hazard Mater 158 (2008) 438–444.
[62] Faisal Khan, Tahir Husain, R. Hejazi, An overview and analysis of site remediation technologies Original Research Article, Journal of Environmental Management, 71 (2004) 95-122.
[63] D.M. Hamby, Site remediationtechniques supporting environmental restoration activities—a review, Science of The Total Environment, 191 (1996) 203-224.
[64] R. Flores, M.G. García, J.M. Peralta-Hernández, A. Hernández-Ramírez, E. Méndez, E. Bustos, Electro-Remediation in The Presence of Ferrous Sulfate as an Ex-Situ Alternative Treatment for Hydrocarbon Polluted Soil, Int. J. Electrochem. Sci.,, 7 (2012) 2230 - 2239.
[65] Dura´n N, Use of nanoparticles in soil-water bioremediation processes., J Soil Sci Plant Nutrit 8(2008) 33–38.
[66] Rajendran P, G. P, Nanotechnology for bioremediation of heavy metals, Environmental bioremediation technologies, (2007) 211–221.
[67] E.L. Davis, Steam Injection for Soil and Aquifer Remediation, United States Environmental Protection, (1998) 1-16.
[68] M.M. Amro, Treatment Techniques of Oil-Contaminated Soil and Water Aquifers, International Conf. on Water Resources & Arid Environment, (2004) 1-11.
[69] L. K, Y. Li, Chemical Reduction/Oxidation Advanced Physicochemical Treatment Processes, in: L.K. Wang, Y.-T. Hung, N.K. Shammas (Eds.), Humana Press, 2006, pp. 483-519.
[70] O.A.P. Ltd, Groundwater Treatment Plant Review of Treatment Options, Final report, (2004) 1-93.
[71] W.E. Schwinkendorf, Evaluation of Alternative Nonflame Technologies for Destruction of Hazardous Organic Waste, (1997) 1-175.
[72] L.L. Tavlarides, W. Zhou, G. Anitescu, SUPERCRITICAL FLUID TECHNOLOGY FOR REMEDIATION OF PCB/PAH-CONTAMINATED SOILS/SEDIMENTS, Proceedings of the 2000 Conference on Hazardous Waste Research, (2000) 1-17.
[73] M.D.A. Saldaña, V. Nagpal, S.E. Guigard, Remediation of Contaminated Soils using Supercritical Fluid Extraction: A Review (1994-2004), Environmental Technology, 26 (2005) 1013-1032.

Chapter 5

Polycyclic Aromatic Hydrocarbons a Constituent of Petroleum: Presence and Influence in the Aquatic Environment

Daniela M. Pampanin and Magne O. Sydnes

Additional information is available at the end of the chapter

http://dx.doi.org/10.5772/48176

1. Introduction

Crude oil is a complex mixture of hydrocarbons containing more than 17000 compounds [1]. Among the constituents of crude oil there is a group of substances called polycyclic aromatic hydrocarbons (PAHs). PAHs are aromatic compounds containing from two to eight conjugated ring systems. They can have a range of substituents such as alkyl, nitro, and amino groups in their structure [2]. Nitrogen, sulfur, and oxygen atoms can also be incorporated into their ring system [2,3]. The precursors for PAHs found in crude oil are natural products, such as steroids, that have been chemically converted to aromatic hydrocarbons over time [4].

The PAHs that are present in the marine environment in relevant concentrations are divided into two groups depending on their origin, namely pyrogenic and petrogenic [5]. Pyrogenic PAHs are formed by incomplete combustion of organic material while the petrogenic PAHs are present in oil and some oil products [4,6,7]. In general the pyrogenic PAHs are composed of larger ring systems then the petrogenic PAHs. Sources for pyrogenic PAHs are forest fires [6,7,8], incomplete combustion of fossil fuels [6,7,8], and tobacco smoke [6,7]. A range of PAHs are naturally present in crude oil [4,9,10] and coal [10,11] and these compounds are referred to as petrogenic PAHs. In the costal zones PAHs enters the water primarily from sewage, runoff from roads [12], the smelter industry [13,14,15] and oil spills [16,17], while offshore PAHs chiefly enter the water through oil seeps [18], oil spills [16], and produced water discharge from offshore oil installations [19].

2. Oil as a source of polycyclic aromatic hydrocarbons to the aquatic environment

Hydrocarbons in the form of crude oil, and therefore also PAHs, have and are entering the environment naturally through oil seeps, which is oil leaking naturally from oil reservoirs. Oil seeps are found scattered all over the globe with a higher concentration in certain regions of the world [18]. Numerous times the presence of a natural oil seep has resulted in the discovery of oil reservoirs that are large enough for commercial oil production [20], however, the presence of an oil seep does not guarantee the discovery of a production worthy oil reservoir [21]. Oil seeps vary in size with macroseeps resulting in visible oil slicks on the water surface, when the oil seep is situated on the seafloor, and microseeps that are invisible at the surface [22]. Estimates for the world-wide seepage rate vary between 0.02-2.0 x 10^6 tons per year with the most realistic estimate being 0.2 x 10^6 tons per year [23]. The presence of natural oil seeps also results in local presence of hydrocarbon-eating microorganisms [24,25], a fact that gives these regions an advantage in the case of an accidental oil spill [24,26]. For example resent research showed that the presence of oil eating microorganisms due to natural oil seeps in the Gulf of Mexico in addition to favorable water currents resulted in a quicker degradation of oil in the region after the *Deepwater Horizon* accident than otherwise would be expected [26,27]. As expected, it is the lighter fractions, viz. short chain alkanes, of the oil that are first degraded by microorganisms [28]. The easily accessible energy source, short chain alkanes, results in an explosion like increase in the number of the oil degrading microbes. After some time these microorganisms also start degrading the more complex molecules such as long chain alkanes and aromatic compounds like PAHs [29]. A range of PAH degrading bugs has been found naturally in the environment [30-34]. These microorganisms have been isolated, sequenced and studied extensively in the laboratory, and their mechanism of degrading PAHs is fairly well understood [29].

For monitoring purposes The US Environmental Protection Agency has made a list of 16 unsubstituted PAHs that are on a priority pollutant list [35]. These PAHs are usually referred to as the EPA 16 PAHs (Figure 1) and are the PAHs most commonly analyzed for. The concentration of these 16 PAHs and other PAHs, and the ratios between the various compounds differ from oil to oil [36,37]. This fact is quite clearly presented in the work of Kerr et al. where they have analyzed the concentration of the prioritized 16 PAHs in 48 crude oils from around the globe (North America, South America, Africa, and Asia) [36]. Their results are summarized in Table 1 and show an enormous variation of the content of the various PAHs in crude oil from different sites. Crude oil from the North Sea is reported to have a PAH concentration of 0.83% [38] while for example crude oil that leaked out of Exxon Valdez (referred to as Exxon Valdez crude oil hereafter) had a PAH content of 1.47% [39].

Naphthalene, one of the 16 EPA PAHs, is present in the highest concentration in crude oil, nevertheless this quantity does not give the full details of the naphthalene content of the oil. In fact, if the content of methylated naphthalene, a normal constituents of crude oil, is also included in the analysis the total naphthalene concentration is much higher as illustrated in Table 2. A similar situation is also found for phenanthrene [38,39] and chrysene [38], while for the less abundant PAHs in crude oil excluding the alkylated derivatives of the parent compounds does not alter the total quantity significantly. In fact the alkylated PAHs found in crude oil have been reported to be more toxic then their unsubstituted congeners [41-43].

Crude oil	48 different crude oils [36]			North Sea [38]	Goliat [40][b]
PAH	Minimum mg/kg oil	Maximum mg/kg oil	Mean mg/kg oil	mg/kg oil	mg/kg oil
Naphthalene	1.2	3700	427	1169	1030
Acenaphthene	0	58	11.1	18	12
Acenaphthylene	0	0	0	11	*
Fluorene	1.4	380	70.34	265	75
Anthracene	0	17	4.3	1.5	*
Phenanthrene	0	400	146	238	175
Fluoranthene	0	15	1.98	10	6
Pyrene	0	9.2	-	20	*
Benzo[a]anthracene	0	16	2.88	11	Na[c]
Chrysene	4	120	30.36	26	Na[c]
Benzo[b]fluoranthene	0	14	4.08	4.2	Na[c]
Benzo[k]fluoranthene	0	1.3	0.07	Nd[a]	Na[c]
Benzo[a]pyrene	0	7.7	1.5	1.3	Na[c]
Dibenz[a,h]anthracene	0	7.7	1.25	Nd[a]	Na[c]
Benzo[g,h,j]perylene	0	1.7	0.08	1	Na[c]
Indeno[1,2,3-cd]pyrenec	0	1.7	0.08	Nd[a]	Na[c]

Table 1. Minimum, maximum, and mean PAH content in 48 different crude oils [36], and PAH content in two North European crude oils, North Sea crude oil [38], Goliat crude oil [40] ([a]Nd = not detected; [b]Goliat is situated in the Barents Sea; [c]Na = not analyzed for).

Crude oil	North Sea [38]	Goliat [41]	Exxon Valdez crude oil [39]
PAH	mg/kg oil	mg/kg oil	mg/kg oil
Naphthalene	1169	1030	720
1-Methylnaphthalene	2108	2700	1330
2-Methylnaphthalene	2204	4200	1020
3-Methylnaphthalene	1172	2800	Na[a]

Table 2. Concentration of naphthalene and the three regioisomers of methylnaphthalene in crude oil from three different locations ([a]Na = not analyzed).

PAHs represent the group of compounds in oil that has received the greatest attention due to their carcinogenic and mutagenic properties [38,44]. More precisely, intermediates are formed that are far more toxic than the mother compounds during cellular detoxification of the PAHs *in vivo* [45]. Due to the toxicity of the PAH metabolites the oil industry in some areas of the world (e.g. North Sea, Mediterranean Sea, Australian Northwest Shelf, Gulf of Mexico) are required to monitor their discharges to the aquatic environment [46-49].In the North Sea this is taken care of through the Water Column Monitorin programs [48]. For monitoring purposes the 16 EPA compounds has been chosen as the most important PAHs to analyze for.

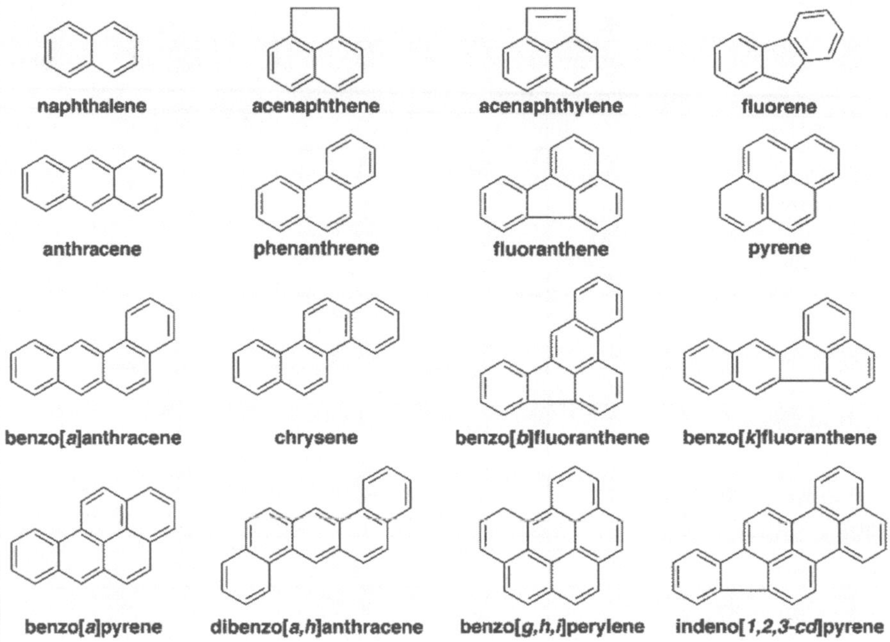

Figure 1. Chemical structure of the EPA selected 16 PAHs.

Unfortunately, accidental oil spills take place from time to time, most recently the *Deapwater Horizon* accident in the Mexican Gulf in 2010 (April 20th). This incident resulted in the release of 779 million liters of crude oil to the Gulf [50]. The *Exxon Valdez* oil spill in Prince William Sound, Alaska, USA in 1989 (March 24th), which took place after the oil tanker ran ashore, resulted in the release of 42 million liters of crude oil [16]. In European waters the *Erika* accident in 1999 (December 12th) released 18000 tons of crude oil into French coastal waters and in Spain *Prestige* spilt 60000 tons of heavy fuel oil into the waters outside of Galicia in 2002 (November 13th) [17]. These are unfortunately only a few examples of accidental release of large quantities of oil to the aquatic environment. The environmental consequences of the *Exxon Valdez* spill is the most studied oil spill ever [51,52], however, the influence of the *Deapwater Horizon* oil release will probably be just as well studied, or even more studied [53].

3. Produced water as a source of polycyclic aromatic hydrocarbons to the aquatic environment

Oil production offshore (Figure 2) (and onshore) also results in the production of large volumes of water, so called produced water (PW), in addition to crude oil. PAHs contained in

PW are receiving much attention due to their potential for causing adverse effects in the marine environment [54,55]. PW generally consists of a mixture of: 1) formation water contained naturally in the reservoir; 2) injected water used for the recovery of oil; and 3) treatment chemicals added during production [19]. Data from offshore oil production platforms in the North Sea have showed that the major aromatic compounds in PW are BTEX (benzene, toluene, ethylbenzene and xylene) (97%), 2- and 3-ring PAHs (3%) named NPD (naphthalenes, phenanthrenes and dibenzothiophenes) and larger PAHs (<0.2%) [19,56,57].

The average concentrations of NDPs and the remaining 14 PAHs from the EPA 16 list are shown in Table 3. With regard to the NPD group, the lowest molecular weight substance, naphthalene represents the largest fraction (92%). Even PAHs with more than 3-rings constitute less than 0.2% of the aromatic content of PW, with the concentrations of the individual compounds decreasing with the number of rings in their structure, by about one order of magnitude for each additional ring.

16 PAHs from the EPA list	Oil installation	Gas installation	Unspecified installation
Naphthalene	145	115	108
Phenanthrene	13.6	20.9	10.7
Fluorene	8.3	13.1	6.7
Acenaphthene	2	50.1	1.78
Acenaphthylene	0.86	12.6	2.35
Fluoranthene	0.26	35.4	0.29
Anthracene	3.74	110	1.17
Pyrene	0.63	8	0.47
Benzo[a]pyrene	0.52	-	0.022
Chrysene	0.84	1	0.52
Benzo[a]anthracene	0.23	1	0.25
Benzo[b]fluoranthene	0.028	-	0.031
Benzo[k]fluoranthene	0.007	-	0.007
Dibenz[a,h]anthracene	0.005	-	0.005
Benzo[g,h,j]perylene	0.029	-	0.019
Indeno[1,2,3-cd]pyrenec	0.005	-	0.006

Table 3. Mean concentrations of aromatic compounds in produced water (expressed as μg/L) (data from OLF, UKOOA and OGP companies, [58]).

PAHs in PW are present in both dissolved and dispersed (oil droplets) forms. Existing oil/water separators, such as hydrocyclones, are efficient in removing droplets, but not dissolved hydrocarbons from PW. Therefore, much of the PAHs discharged in to the marine environment are dissolved low molecular weight compounds. Nevertheless, since the treatment procedures are not 100% effective, discharged PW still contains some dispersed oil (droplet size from 1 to 10 μm) [59]. Concentration of total PAHs in PW typically range from 0.040 to 3 mg/L (Table 4) and consist primarily of the most water-soluble congeners such as naphthalene, phenanthrene and their alkylated homologues (2- and 3-ring PAHs). The abundance of these alkyl substituted PAHs is also higher than for the parent compounds (the non-alkylated homologues). Higher molecular weight PAHs (up to 6-ring) are in fact rarely detected in properly treated PW [60,61]. The ratio between alkylated-PAHs and the corresponding parent compounds is therefore used to confirm the nature of the pollution source in field studies [54,62].

Due to their lipophilic properties, PAHs are mainly associated with dispersed oil droplets [59,63]. In fact, it has been documented that up to 10% of the total PAHs in PW from a platform on the Northwest Shelf of Australia were in the dissolved fraction being formed mainly by alkylnaphthalenes and alkylphenanthranes [64]. Moreover, these droplets also contained almost all the dibenzothiophenes, fluoranthenes/pyrenes and chrysenes present in the PW.

Figure 2. The sources of PAHs entering the marine environment offshore are predominantly natural oil seeps, oil spills from boats or platforms, and produced water discharge from oil and gas producing installations such as the one shown. PAHs in produced water and oil seeps represent a chronic release to the marine environment.

As a chronic source of PAH contamination in the marine environment, PW is also a source of concern with respect to possible long term impact on the environment [65]. Estimates of the PW discharge volumes on the Norwegian shelf predict an increase until 2010–2014, reaching a maximum of about 200 million L/year [66]. Therefore, this offshore discharge is currently under periodical monitoring [48,67].

Compound	Gulf of Mexico	North Sea	Scotia Shelf	Grands Bank (Canada)
Naphthalene	5.3-90.2	237-394	1512	131
C_1-Naphtalenes	4.2-73.2	123-354	499	186
C_2-Naphtalenes	4.4-88.2	26.1-260	92	163
C_3-Naphtalenes	2.8-82.6	19.3-81.3	17	97.2
C_4-Naphtalenes	1.0-52.4	1.1-75.7	3.0	54.1
Acenaphthylene	ND-1.1	ND	1.3	2.3
Acenaphthene	ND-0.1	0.37-4.1	ND	ND
Biphenyl	0.36-10.6	12.1-51.7	ND	ND
Fluorene	0.06-2.8	2.6-21.7	13	16.5
C_1-Fluorene	0.09-8.7	1.1-27.3	3	23.7
C_2-Fluorene	0.20-15.5	0.54-33.2	0.35	4.8
C_3-Fluorene	0.27-17.6	0.30-25.5	ND	ND
Anthracene	ND-0.45	ND	0.26	ND
Phenanthrene	0.11-8.8	1.3-32.0	4	29.3
C_1-Phenanthrene	0.24-25.1	0.86-51.9	1.30	45.0
C_2-Phenanthrene	0.25-31.2	0.41-51.8	0.55	37.1
C_3-Phenanthrene	ND-22.5	0.20-34.3	0.37	24.4
C_4-Phenanthrene	ND-11.3	0.50-27.2	ND	13.2
Fluoranthene	ND-0.12	0.01-1.1	0.39	0.51
Pyrene	0.01-0.29	0.03-1.9	0.36	0.94
C_1-Fluoranthene/Pyrenes	ND-2.4	0.07-10.3	0.43	5.8
C_2-Fluoranthene/Pyrenes	ND-4.4	0.21-11.6	ND	9.1
Benzo[a]anthracene	ND-0.20	0.01-0.74	0.32	0.60
Chrysene	ND-0.85	0.02-2.4	ND	3.6
C_1-Chrysenes	ND-2.4	0.06-4.4	ND	6.3
C_2-Chrysenes	ND-3.5	1.3-5.9	ND	18.8
C_3-Chrysenes	ND-3.3	0.68-3.5	ND	6.7
C_4-Chrysenes	ND-2.6	ND	ND	4.2
Benzo[b]fluoranthene	ND-0.03	0.01-0.54	ND	0.61
Benzo[k]fluoranthene	ND-0.07	0.006-0.15	ND	ND
Benzo[e]pyrene	ND-0.10	0.01-0.82	ND	0.83
Benzo[a]pyrene	ND-0.09	0.01-0.41	ND	0.38
Pyrelene	0.04-2.0	0.005-0.11	ND	ND
Indeno[1,2,3-cd]pyrenec	ND-0.01	0.022-0.23	ND	ND
Dibenz[a,h]anthracene	ND-0.02	0.012-0.10	ND	0.21
Benzo[g,h,j]perylene	ND-0.03	0.01-0.28	ND	0.17
Total PAHs	40-600	419-1559	2148	845

Table 4. Concentrations of individual PAHs or alky congener groups in produced water from various areas (expressed as μg/L) (ND = not detected) (from [61]).

4. Environmental monitoring of polycyclic aromatic hydrocarbons

There are various environmental monitoring methods which may be performed in order to assess risks of PAH contamination for organisms and to classify the environmental quality of ecosystems. Five approaches are hereby reported: 1) chemical monitoring: exposure assessment by measuring levels of a selected set of compounds in abiotic environmental compartments; 2) bioaccumulation monitoring: exposure assessment by measuring PAH levels in biota or determining the critical dose at a critical site (bioaccumulation); 3) biological effect monitoring: exposure and effect assessment by determining the early adverse alterations that are partly or fully reversible (biomarkers); 4) health monitoring: effect assessment by examining the occurrence of irreversible diseases or tissue damage in organisms; 5) ecosystem monitoring: assessment of the integrity of an ecosystem by making an inventory of, for instance, species composition, density and diversity [68-70]. All these methods are currently in use to monitor the aquatic environment contamination from PAH compounds.

Since the occurrence and abundance of PAHs in aquatic environments represent a risk to aquatic organisms and ultimately to humans (through fish and shellfish consumption), there is a constant need for their determination and quantification around the world [71]. The monitoring of PAH presence in the aquatic environment is therefore a world-wide activity. Since some of these compounds are well known carcinogens and mutagens [44,72], this contaminant class has been generally regarded as high priority for environmental pollution monitoring. In fact, the European Union included these pollutants in the list of priority hazardous substances for surface waters in the Water Framework Directive 2000/60/EC [73]. Moreover, for PAH content in biota several guidelines exist: the commission regulation (EC 2005) which stipulates a maximum concentration of 10 μg/kg (w/w) of benzo[*a*]pyrene in edible molluscs (this regulation also limits the benzo[*a*]pyrene concentration in other alimentary products), the OSPAR Commission which has developed eco-toxicological assessment criteria (i.e., concentrations levels above which concern is indicated) for different PAHs in fish and mussels [74], and the Oregon Health Division which has derived risk-based criteria for PAHs [75] assessing the risk in terms of benzo[*a*]pyrene equivalents.

Many studies have been carried out to determine the PAH distribution in marine organisms in different geographical areas using different approaches. Different authors have reported the presence of PAHs in waters, marine organisms, and sediments using chemical and biological markers [76-78]. Current monitoring techniques employed to determine environmental quality include the chemical analyses of sediment and water samples for determining the concentration of PAH parent compounds. PAHs are sparingly soluble in water and are difficult to detect although they show a much greater association with sediments. In fact, due to their hydrophobic character, these compounds rapidly tend to become associated with particles and end up in the sediments which act as a sink for them [71,79,80]. High concentrations of pyrogenic PAH mixtures in sediment samples have been found in several freshwater, estuarine and marine regions with heavy vessel traffic or at locations with PAH-containing effluents from industrial areas [81]. Finally, chemical analyses of these materials are laborious and costly due to the lengthy extraction methodology needed. Since monitoring of a large number of PAHs is expensive, time consuming and analytically demanding, a selection of compounds could be monitored to give an indication of the overall contamination.

Standard procedure for determining PAHs in the aquatic environment have a long history that covers at least three decades. A few examples are reported here (see also Table 5). One of the first official procedures was published in 1985 by HMSO (Her Majesty's Stationery Office) and provided the analysis of six PAHs in water samples [82]. In 1986, the EPA published EPA procedure 8310 for determining PAHs by HPLC [83]. In 2007, the EPA published the EPA procedure 8270D for determination of 16 PAHs applying gas chromatography coupled with mass spectrometry (GC-MS) [84]. To close, the standard method for PAH determination in water recommended by the International Organization for Standardization (ISO), ISO 17993:2002, is based on liquid-liquid extraction and final determination by HPLC [85].

Method name	Application	Sample preparation	Determination technique
EPA-610	PAHs in municipal and industrial wastewaters	About 1L of water samples is extracted with dichloromethane, the extract is then dried and concentrated to a final volume of less than 10 mL.	HPLC or GC (with a packed column)
EPA-8310	Groundwater and wastewater	EPA 3500 series methods (version 3, 2000)	HPLC
EPA-8270D	Aqueous samples	EPA 3500 series methods (version 3, 2000)	GC-MS with use of high resolution capillary columns and deuterated internal standard
ISO 17993:2002	Water quality (determination of 15 PAHs)	Water samples collected in brown glass bottles are stabilized by adding sodium thiosulfate. 1 L of sample is extracted using hexane. The extract is dried with sodium sulfate and then enriched by removal of hexane by rotary evaporation	HPLC and fluorescence detector

Table 5. Standard procedures for determining of PAHs in aqueous samples (modified from [86]).

Modern analyses have at their disposal a wide spectrum of tools, specialized analytical equipment and standard procedures that are capable of providing high quality results even if intercalibration studies and commonly accepted SOP are needed [86].

Chemical fingerprinting of a PAH mixture can lead to the identification of the contamination source, for example an oil spill accident [87]. PAH compounds in a petrogenic exposure often contain one or more methyl- ethyl- or butyl-(and sometimes higher alkyl-) groups on one or more of the aromatic carbons [88]. For example, the ratio between alkyl-PAH compounds and the corresponding parent compounds is commonly used to confirm the nature of the pollution source in field studies from PW discharges (i.e. the abundance of alkyl substituted PAHs is higher than for the parent compounds (the non-alkylated homologues)). It has been shown that mussels caged down-stream of PW discharges from oil platforms accumulate higher concentrations of alkylnaphthalenes, alkylphenanthrenes and alkyldibenzothiophenes, than their respective parent compounds [62]. In a recent study, the ratio of alkylated over non-alkylated PAHs indicated a diffuse petrogenic contamination in the Ekofisk area (up to 2000 m from PW fallout) [55].

PAHs co-occur in various amounts and the composition of the mixture differs depending upon the source from which they are derived and their subsequent degradation. For example a recent study clearly demonstrated the correlation between PAH contamination in river water and the nearby activity of textile factories [89]. A widespread PAH contamination in the San Francisco Bay has been reported after a 10 year monitoring of water, sediment and biota samples [90]. Since the process of industrialization and urbanization is growing rapidly in South America, the potential increase in PAH contamination is under evaluation through a number of surveys in the coastal areas, collecting information about the PAH concentration in water, sediment and biota [91-93]. Of course, oil spills are well-known examples of PAH contamination in the aquatic environment. Unfortunately, many cases of studies are reported in the literature from different parts of the world, such as the Gulf of Mexico in [94], the Mediterranean Sea [95], the Spanish coasts [96], and the Philippines islands [97].

PAH contamination is also constantly under the attention of different countries in relation to oil and gas explorations [98]. For example the Water Column Monitoring program financed by Oljeindustriens Landsforbund (OLF), has provided information about PAH contamination from platform discharges in the North Sea since 2001, almost yearly based [48]. This monitoring program is a good example of integration of chemical monitoring methods with biological effect monitoring approaches in PAH monitoring.

Passive sampling devices, such as semipermeable membrane devices (SPMDs) and polar organic chemical integrative samplers (POCIS) also provide a useful contribution to the monitoring of PAH contaminants in the aquatic environment [99-101]. The principle of the passive sampling technique is the placement of a device in the environment for a fixed period of time, where it is left unattended to accumulate contaminants by diffusive and/or sorptive processes. The main advantages of using passive sampling devices over traditional discreet spot water samples are: 1) concentrations are time-integrative during exposure, compensating for fluctuations in discharges; 2) lower detection limits are normally achievable as a larger sample has been taken and; 3) only the freely dissolved and thus more readily bioavailable fraction is measured. Furthermore, there is a reduction in the need for the use of animals in scientific experiments. However, relating passive sampling device accumulations to the overall ecological relevance of contaminants is complicated and effects can only be inferred.

Nevertheless, the chemical approach does not provide information about PAH bioavailability, and the toxic potential and the risk posed by the potentially much more toxic daughter compounds produced by many organisms as a result of metabolism and biotransformation of the parent chemicals is not taken into account [102]. Therefore, a biological effect monitoring approach using living organisms for PAH environmental monitoring has been developed during the last couple of decades. Metabolites arising from biotransformation processes maybe concentrated in body fluids, tissues or excreta and the analysis of such biological compartments provides an opportunity to detect and measure exposure of organisms to bioavailable contaminants [103]. On the other hand, PAHs are known to induce toxic effects at the individual level [104,105], and integration of chemical analyses with biomarker responses in organisms has been recommended for monitoring of PAH contamination, in particular in oil related activities (e.g. offshore exploitation activities) [48]. The feasibility of using tissue concentrations of PAH compounds in marine species as a marker of environmental contamination depends on the relative rates of uptake, biotransformation and excretion of the organism. Invertebrate filter feeders, such as *Mytilus* spp., are highly efficient accumulators and bioconcentrators of PAHs and therefore commonly used [54,106-108]. Total amounts of the EPA 16 PAHs between 10 μg/kg and 20 μg/kg, have been found in mussels caged in the vicinity of Norwegian platforms [48,54,109]. Fish and other invertebrates, rapidly biotransform PAHs and their presence in tissues is low and no representative of the overall contamination. PAHs related to offshore operational discharges in fact are generally not found in muscle of wild specimens of fish collected in regions with oil and gas activity [110].

Some biomarkers are widely used as sensitive and early warning signals of exposure to PAHs. For example, many studies indicated that PAH compounds were detectable several kilometers away from North Sea oil production platforms using *in vitro* bioassays and biomarkers [111]. Currently the induction of ethoxyresorufin-O-deethylase (EROD) activity, the production of bile metabolites and the formation of DNA adducts have shown the greatest potential for identifying level of exposure to PAHs following contamination of the aquatic environment [87,112-114]. The historically commonly used marker is the induction of Cytochrome P450 1A (CYP1A) in fish measured by the catalyzed O-deethyulation of ethoxyresorufin in hepatic microsomes [104,115-118]. The induction of CYP1A in fish following exposure to certain classes of organic contaminants has been the basis of the use of the cytochrome P450 system as a biomarker in pollution monitoring since the 70's [104]. Many studies reported the used of this biomarker in various monitoring surveys since the '80s [104,119,120]. In many cases, EROD activities or CYP1A protein levels were correlated with environmental levels of CYP1A-inducing chemicals such as PAHs. Nevertheless, a linear dose–response relationship cannot always be found between the PAH concentration and the CYP1A content and/or activity in the natural environment, where a mixture of both inducers and inhibitors of CYP1A may act simultaneously [121]. Moreover, other factors (e.g. temperature, season or sexual hormones) can also modulate the responsiveness of the

CYP1A system in fish [122]. Since PAH compounds are absorbed via gills and may be metabolized before reaching the liver, hepatic EROD activity may not be the only and most sensitive organ to reflect the presence of CYP1A inducing agents (such as PAHs) in water. Therefore, a sensitive method to determine EROD activity in gill filaments has also been developed [123].

A very efficient tool to assess PAH exposure in fish is the determination of PAH metabolites in fish bile. They can be measured using several analytical methods from the simple and fast fluorescence assay (fixed fluorescence detection or synchronous fluorescence spectrometry) to the HPLC with fluorescence detection (HPLC-F) after deconjugation, extraction and derivatization of the bile samples, to the extremely sensitive and advanced LC-MS/MS and GC-MS/MS methods. These methods are very different both in regard to their analytical performances towards different PAH metabolites as well as in technical demands and monitoring strategies. A recent review reported the state of the art for the different methods for determining metabolites of PAH pollutants in fish bile [87]. This approach has also been developed for crustaceans. Metabolites in crabs urine have been analyzed to monitor environmental contamination from PAH with success [103,124]. Regarding DNA adduct analysis as a biomarker of exposure to PAHs, a description of methods is reported in the section DNA adducts from polycyclic aromatic hydrocarbons.

5. Polycyclic aromatic hydrocarbon metabolites

The strongly hydrophobic PAHs accumulate in fatty tissue such as liver, where they penetrate the cells by means of passive diffusion. Inside the hepatocytes, PAHs are oxidized and hence made more water-soluble and more reactive by enzymes with aryl hydrocarbon hydroxylase activity to form epoxides and diols according to the general route outlined in Figure 3 [125]. CYP1A is the most important and best described aryl hydrocarbon hydroxylase enzyme participating in the first step (Phase I) of xenobiotic detoxification [126,127]. CYP1A-derived metabolites generally have a high affinity to nucleic acids and proteins, which may result in adduct formation and possible impaired function of these biomolecules.

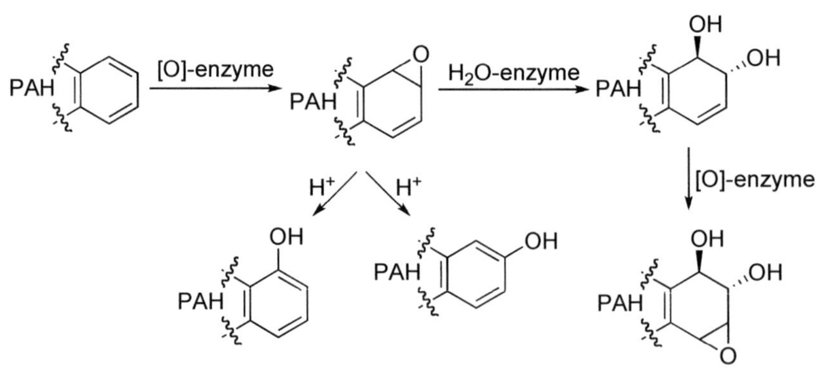

Figure 3. General outline of the metabolic degradation of PAHs.

The vast majority of Phase I-derived PAH metabolites are passed on to Phase II (most notably glutathione transferase) and Phase III (most notably the transmembrane ATP-binding cassette exporter proteins) to eventually become excreted with the bile fluid [126]. In this context, the presence of PAH metabolites in the bile of fish is a highly regarded biomarker of recent PAH exposure (exposure that has taken place within a few days prior to sampling and analysis) (*vide supra*) [37,87].

The various PAHs form a range of metabolites *in vivo*. Most of the studies mapping these metabolites have been performed on humans, rats, mice, and hamsters [128]. However, some studies have been conducted on fish and/or fish cells. The results from these studies show that the point of oxidation varies from species to species for the same PAH due to the presence of various cytochrome P450 (CYP) isoforms [128]. For fish it is predominantly CYP1A's role in PAH metabolism that has been studied, however, for humans, in particular, other isoforms of CYP have also been well investigated. For a general discussion regarding the various CYP families found in fish, see Uno *et al.* [129]. Figure 4 summarizes the data reported for oxidation site for a range of PAHs, phenanthrene [130,131], chrysene [131,132], pyrene [130,133], benz[*a*]pyrene [131,134], benzo[*c*]phenanthrene [135,136], and dibenzo[*a,l*]pyrene [135-137], based on *in vitro* tests with fish CYP1A. Most of the PAHs shown in Figure 4 have predominantly one major site where oxidation takes place, thus, indicating a high regioselectivity in the enzymatic oxidation by CYP.

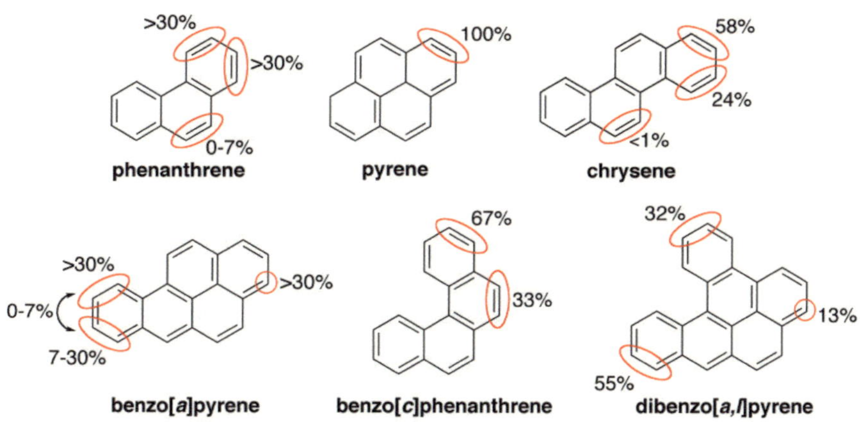

Figure 4. Point of oxidation with distribution intervals.

The absolute stereochemistry of PAH metabolites, derived from metabolism of PAHs in the liver, influences the toxicity of the metabolite [138,139]. In particular the diols with *R,R*-configuration and the *R,S*-diol-*S,R*-epoxides show high carcinogenic activity [139]. As shown in Figure 4 for chrysene the predominant diol formed is the 1,2-diol (formed in 58%), with 3,4- and 5,6-diol being formed in 24% and >1%, respectively (structures for the diols are shown in Figure 5) [132]. The data for chrysene depicted in Figure 4 were based on *in vitro* tests utilizing liver microsomes from brown bullhead, however, these findings were later confirmed by Jonsson et al. in *in vivo* tests with Atlantic cod [140]. Close to 90% of the chrysene 1,2-diol is formed with the *R,R*-configuration (structure shown in Figure 5) and slightly more than 10% is formed with the *S,S*-configuration [132]. For the 3,4-diol 97% is formed with the *R,R*-configuration (structure shown in Figure 5) [132], thus indicating that in the fish brown bullhead predominantly the most toxic metabolites are formed. The diols derived from chrysene are only considered to have weak carcinogenic activity [141-143], however, 1,2-dihydroxy-1,2-dihydrochrysene is the starting point for the biosynthesis of the most carcinogenic chrysene metabolite 1,2-dihydroxy-3,4-epoxy-1,2,3,4-tetrahydrochrysene [144].

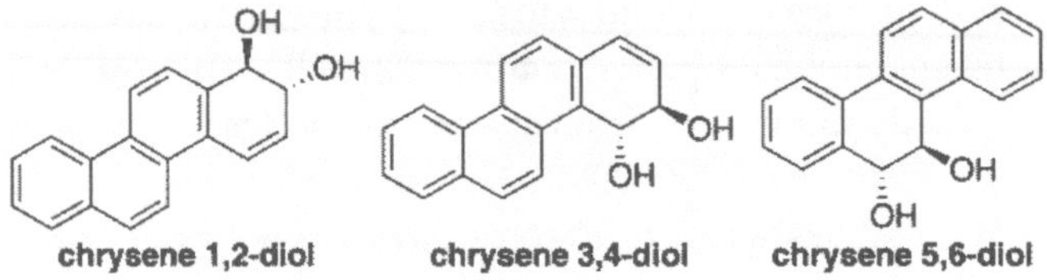

Figure 5. Chemical structure of chrysene metabolites.

Figure 6 and 7 summarizes the structure of the metabolites formed by *in vivo* oxidation of benzo[*c*]phenanthrene [145], and benzo[α]pyrene, respectively [146,147]. The two major metabolites formed from benzo[*c*]phenanthrene are compounds **1** and **2**. Biological testing revealed that metabolite **3** from benzo[*a*]pyrene was not carcinogenic while compound **4** was carcinogenic (Figure 7) [146,147]. The carcinogenic metabolite was also the major compounds formed biosynthetically.

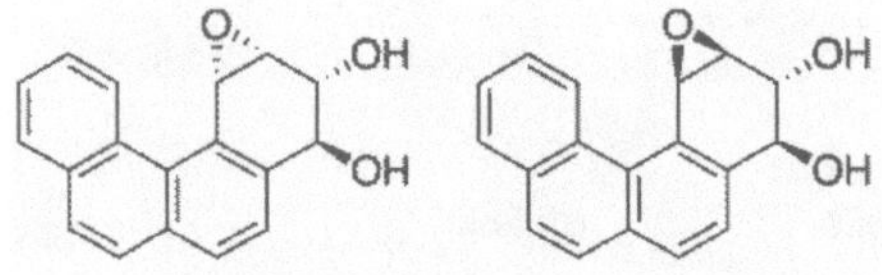

1 2

Figure 6. Chemical structure of benzo[*c*]phenanthrene metabolites.

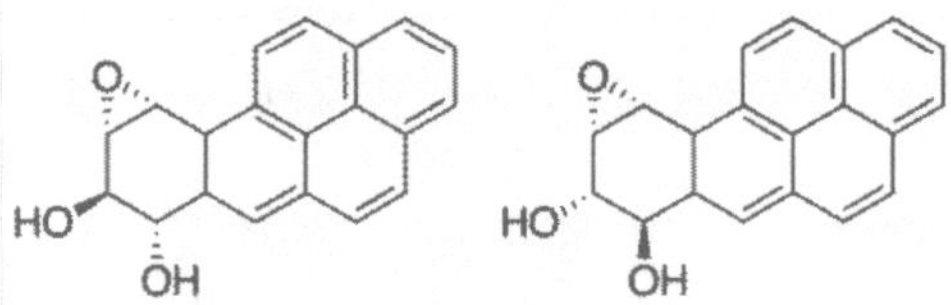

3 4

Figure 7. Chemical structure of 7,8-diol-9,10-epoxide-benzo[*a*]pyrene.

In vivo oxidation in rodents of dibenzo[*a,l*]pyrene, which is considered as the most potent carcinogenic PAHs, results in a range of metabolites as outlined in Figure 8 [148]. The study showed that predominantly the (-)-11*R*,12*R*-enantiomer was formed and that the genotoxic events mainly took place by stereoselective activation of that enantiomer.

The PAHs bay region diol epoxide has been singled out as the cause for this group of compounds carcinogenic activity [149-151]. The diol epoxides have been found to react with cellular macromolecules of paramount importance, namely DNA and proteins [152-154].

Miller proposed in 1970 that PAH metabolites are electrophiles that react with nucleophiles *in vivo*, thus delivering their biological effect [155]. Later it has been proposed that mechanistically the diol epoxides are electrophiles that alkylate the purine bases in DNA via an SN1-like epoxide ring opening process [156,157].

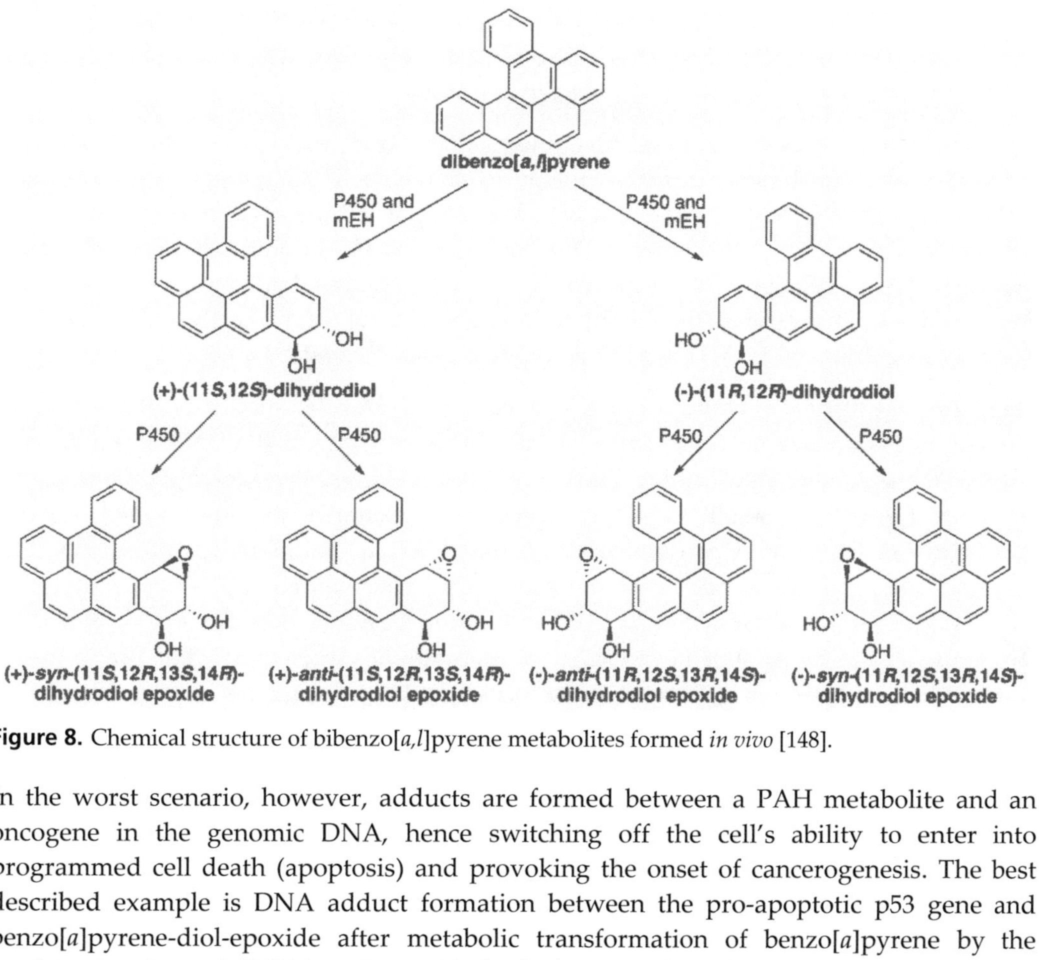

Figure 8. Chemical structure of bibenzo[*a,l*]pyrene metabolites formed *in vivo* [148].

In the worst scenario, however, adducts are formed between a PAH metabolite and an oncogene in the genomic DNA, hence switching off the cell's ability to enter into programmed cell death (apoptosis) and provoking the onset of cancerogenesis. The best described example is DNA adduct formation between the pro-apoptotic p53 gene and benzo[*a*]pyrene-diol-epoxide after metabolic transformation of benzo[*a*]pyrene by the combined actions of CYP1A and epoxide hydrolase, another Phase I enzyme (this is also the mechanism by which cigarette smokers develop lung cancer). The presence of DNA adducts in liver tissue is, unlike CYP1A-induction or accumulation of PAH metabolites in bile, the result of cumulative exposure over weeks or even months [158,159]. Hence, hepatocytic CYP1A induction, accumulation of PAH metabolites in bile and elevated liver DNA adducts represent a chain of events that, although partly separated in time, are tightly interrelated from a mechanistic point of view. These three biomarkers have, therefore, received much attention and represent valuable biomarkers of PAH-exposure and effect in fish [159-162]. Analogous with several studies on mammals, responses of these core biomarkers of oil exposure have been associated with genotoxic effects such as liver neoplasia [163].

6. DNA adducts from polycyclic aromatic hydrocarbons

The toxicity of PAHs is a continuous subject of intense investigation. The carcinogenic potential of PAHs was recognized as early as 1933 by Cook et al. [164], who isolated a cancer-producing hydrocarbon from coal tar. Many PAHs act as potent carcinogens and/or mutagens via DNA adduct formation. Aquatic vertebrates such as fish are capable of metabolizing PAHs (*vide supra*), producing reactive intermediates, occasionally with the formation of hydrophobic DNA adducts as an end result.

Metabolic activation of PAHs to reactive intermediates is a prominent mode of their toxic action. Xue and Warshawsky described the principal metabolic pathways that yield reactive PAH intermediates [165]. Two pathways in particular produce electrophiles that may covalently bind to DNA (forming a DNA adduct): 1) electrophilic diol-epoxides from sequential PAH oxidation by cytochrome P450 (CYP) enzymes, hydrolysis of the resulting arene oxides by microsomal epoxide hydrolase, and a second CYP-catalyzed oxidation; 2) one electron oxidation of PAHs by CYP peroxidase yields the radical anion.

There are 18 potential sites for adduct formation in DNA. The specificity of reactions at different sites depends on the reactive species, nucleophilicity of the DNA site and steric factors. The spectra of DNA adducts resulting from PAHs are considerably different from the ones formed by small alkylating agents. For example the dihydrodiol epoxide metabolites of PAHs react predominantly at the exocyclic amino groups of guanine and adenine. The major DNA adduct of the carcinogen, benzo[*a*]pyrene-7,8-dihydrodiol-9,10-oxide, occurs at N^2 of guanine [166]. DNA binding basically depends on its molecular structure and functional state (accessibility of nucleophilic target sites) while physiological and biochemical features determine differences in adduct formation among tissues and across species.

DNA adducts caused by PAH metabolites are known to be crucial factors in the aetiology of cancer development. For this reason their presence and formation has been profusely studied. Adducts with benzo[*a*]pyrene (recognized as a model compound for the PAH group) are the most frequently reported [167,168], but adducts can also be formed with low molecular weight PAHs, like chrysene, which is a constituent in most mineral oils [169]. A schematic outline of the adduct formation process is outlined in Figure 9.

DNA adducts have been used as a biomarker of exposure to PAHs since the '90s. They represent a very important endpoint, being a marker of genotoxicity. Numerous monitoring studies have reported the formation of DNA adduct formation in aquatic organisms (e.g. fish and bivalves) due to exposure to PAHs. For example, Lyons et al [171] and Harvey et al. [172] reported the genotoxic impact of the *Sea Empress* oil spill accident on different fish species (*Lipophorys pholis* and *Limanda limanda*) as well as on invertebrates (*Halichondira panicea* and *Mytilus edulis*). The DNA adduct patterns of fish liver exhibited the typical diagonal radioactive zone (DRZ) even 17 months after the spill took place. Detection of DNA adducts has been used to assess the impact of the *Erika* oil spill along the coasts of French Brittany. To confirm that the DNA adducts were really related to the *Erika* petroleum, human hepatocyte (HePG2 cells) were exposed to an *Erika* fuel extract. Incubation of HePG2 cells lead to the formation of DNA adducts with similar patterns to the ones observed in the monitoring study using fish (*Solea solea*). These data indicates that human hepatocytes biotransform *Erika* fuel into genotoxic metabolites similarly to hepatic cells of fish and confirmed that the adducts observed in the monitoring study were related to the contamination of the sediment by the oil spill [173].

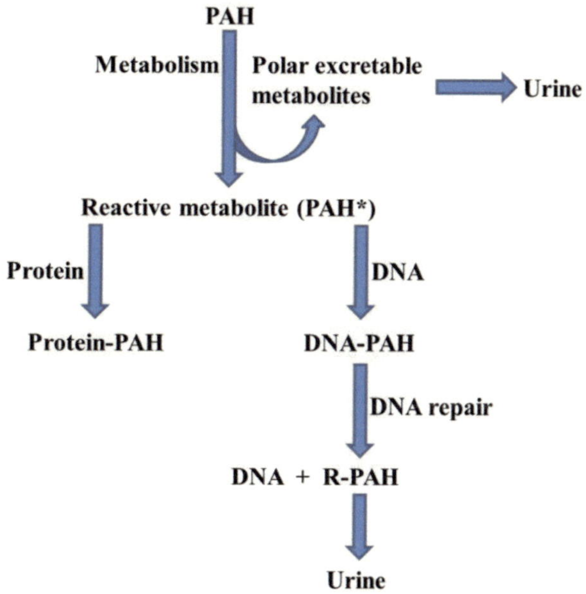

Figure 9. Metabolism of PAH leading to protein and DNA adducts. Figure adapted from reference [170].

It has been shown that DNA adducts persisted in vertebrates species due to the low efficiency of repair systems, representing a parameter for long term exposure [174]. These adducts are very persistent in fish liver [175-178]. French et al. [174] observed a steady increase in DNA adduct levels during a chronic exposure of sole (*Pleuronectes vetulus*) to PAH contaminated sediment for 5 weeks, which were persistent even after a depuration period. The persistency of DNA adducts has been demonstrated also in Atlantic cod (*Gadus morhua*) [38]. In this study, hepatic DNA adducts appeared after 3 days of exposure to low concentration of crude oil (0.06 ppm) and increased steadily during the entire exposure period of 30 days.

Several techniques (e.g. immunoassay, fluorescence assay, gas chromatography-mass spectroscopy (GC-MS), ^{32}P-postlabelling and mass spectrometry (MS) analysis) have been developed for the analysis of PAH derived DNA adducts. At present, the most sensitive and frequently applied technique in aquatic organisms is the ^{32}P-postlabelling assay [179]. Its high sensitivity is unique and achueves the determination of 1-100 adducts in 10^9 nucleotides [180]

The ^{32}P-postlabelling assay appeared in the early '80s and has been applied with a range of protocols in order to detect DNA adducts produced by known carcinogens and complex mixtures [181]. Briefly, the assay involves DNA purification, digestion to normal and adduct-modified 3'-mononucleotides, removal of normal nucleotides (via enzymatic digestion, solvent extraction or chromatographic methods), ^{32}P-postlabelling at the 5' position of adducted nucleotides followed by chromatographic separation, detection and quantification (via autoradiography, scintillation counting or phosphor screen imaging analysis) [182]. Following this assay, PAHs cause the appearance of the diagonal radioactive zone (DRZ) (Figure 10) [173,178,183,184].

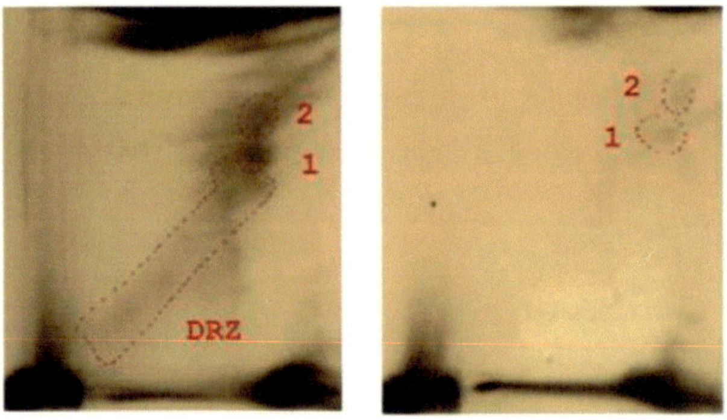

Figure 10. Example of the bulky DNA adduct zone (DRZ) typical of a contamination by PAHs detected by the ^{32}P-postlabelling method (liver sample of fish collected in a PAH contaminated coastal area) (Pampanin unpublished data).

MS/MS analysis has recently emerged as a powerful tool in the detection and structure elucidation of DNA adducts as well as for their quantification at very low concentrations (as often present in biological samples). In fact, electrospray ionization tandem mass spectrometric (ESI-MS/MS) analysis was capable of revealing DNA adducts in different aquatic organisms (e.g. fish (*Oreochromis mossambicus*) and mussel (*Perna perna*) soft tissue) [185,186]. This MS/MS approach provided a rapid determination and discrimination of structurally different phenanthrene derived DNA adducts in fish bile samples [186]. This technique is able to detect one modified base in 10^6-10^{12} unmodified bases [187].

Development of methodologies to detect DNA damages induced by PAHs is of constant concern in aquatic ecotoxicology. Direct chemical methods, such as high performance liquid chromatography with electro-chemical detection (HPLC-EDC), GC-MS and the ^{32}P-postlabelling, are highly sensitive and specific [180,188], however, they are very time and money consuming. A number of antibodies have been generated against carcinogenic products of DNA modifications, including those generated by PAHs. Immunoassays (immunohistochemistry or ELISA) are routinely employed to detect DNA adducts in humans, while the use of such approaches is more limited in aquatic species [189]. An immunoperoxidase method for revealing 7,8-dihydro-8-oxodeoxyguanosine (8-oxo-dG) in marine organisms has been described [189]. This work was also followed by the use of immunofluorescence and antibodies for DNA adduct detection in both vertebrates (fish,

Anguilla anguilla) and invertebrated (mussel, *Mytilus galloprovincialis*). The immunohistochemical approach demonstrated a good selectivity, low cost, is easy to perform and readily allowed the analysis of a large number of samples. Nevertheless, it does not reach the high sensitivity of other methods [190].

7. Protein adducts from polycyclic aromatic hydrocarbons

As discussed earlier (*vide infra*) CYP1A derived PAH metabolites have a high affinity to nucleic acids and proteins, which may result in adduct formation (Figure 11). It is likely that PAH protein adducts are formed after an initial docking, viz. protein-ligand interaction, of the PAH metabolite to the protein followed by reaction with nucleophilic sites in the vicinity of the docking site [191,192]. This has been shown to be the case for human serum albumin, which is predominantly alkylated at histidine[146] by diol epoxides of fluoranthene and benzo[*a*]pyrene [193]. It is highly likely that the same type of mechanism is operating in animals and fish.

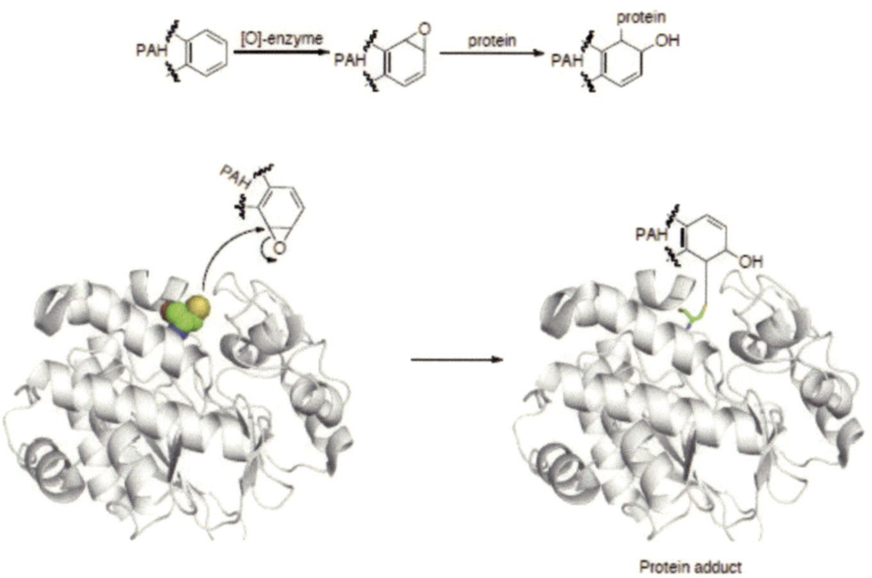

Figure 11. General mechanism for PAH protein adduct formation.

Naturally, adduct formation between PAH metabolites and human proteins and in particular human hemoglobin and serum albumin have been very well documented [170,192-199]. However, there are a few studies concerning protein adducts in fish. Plasma albumin adducts have been isolated from two species of fish (Brook trout (*Salvelinus fantinalis*) and Arctic charr (*Salvelinus alpinus*)) [200]. In the studies conducted by Padrós and Pelletier, which were conducted by injection of benzo[*a*]pyrene, it was found that only the (+)-*anti*-diol-epoxide of benzo[*a*]pyrene generated adducts with serum albumin. In that study it was also found that there was no accumulation of the adduct upon repeated injections, thus indicating a relatively short half-life of less than two days for serum albumin in fish. In humans the half-life of serum albumin has been reported to 20 days [201]. In fish the presence of this adduct would be an indication of a very recent exposure to benzo[*a*]pyrene, while in humans this would also function as a marker of longer term exposure to the PAH. It has been found that the point of adduct formation between human serum albumin and benzo[*a*]pyrene *anti*-diol epoxide is dependent on the stereochemistry of the PAH metabolite [194]. The (+)-enantiomer generates a carboxylic ester adduct with Asp^{187} or Glu^{188} and that the (-)-enantiomer forms an adduct with His^{146}.

The different isoforms of hemoglobin present in different species results in the formation of different adducts. For example, rat hemoglobin possesses a reactive β-cysteine in position 125 not present in human hemoglobin [202]. *syn* And *anti* fluoranthene diol-epoxides form adducts with this particular cysteine in rats. The presence of different isoforms of the same protein in different species results in the possibility of generating different adducts for the same PAH in the different species. The point of adductation most likely reflects on the proteins ability to function. Thus, the adduct formation of a specific protein might affect one species more severely than another.

8. Future perspective

Environmental research related to PAHs has to date, with a few exceptions, predominantly been concerned with finding metabolites of the compounds and detecting the presence of PAH DNA adducts. However, based on the discussion herein it is clear that the next step has to be towards analysis that can provide clear answers regarding the stereochemical outcome of the oxidation processes taking place *in vivo*. By such means it is easier to evaluate the toxicity of the various PAHs. This is a rather large task since different species metabolize PAHs differently resulting in dissimilar distributions between the stereoisomers.

PAH protein adducts have been studied extensively for humans (*vide supra*) and rodents, however, for fish and other aquatic animals this topic is barely touched upon. Studies of adduct formation in fish will further aid the evaluation of the toxicity of the different PAHs. In the PAH protein adduct studies that have been conducted on other species we have seen that the point where the adduct is formed in a specific protein varies from species to species. Generating new knowledge as to where adducts are formed with the same PAH in the same protein in other species might in addition to providing increased knowledge as to the impact of adduct formation also possibly generate new interesting research questions. Hemoglobin in fish has a very short half-life so adduct formation on hemoglobin might not have such a great health impact on the fish. However, other less abundant proteins in the blood are also most likely susceptible to adduct formation with PAH metabolites. Detecting the proteins affected and determining the site of adduct formation will aid in the overall judgment of the toxicity of the PAH responsible for the adduct formed. Adducts with proteins present in the bile may also be of value in assessing the toxicity of PAHs.

In human health care proteomics has for some time been extensively used for diagnostics [203,204] and these techniques are also slowly making their way into ecotoxicology [205,206]. It has been found in human amniotic epithelial cells exposed to anti-7,8-

dihydroxy-9,10-epoxy-7,8,9,10-tetrahydrobenzo[*a*]pyrene, a compound that causes adduct formation on DNA and oxidative damage on DNA, resulting in alternations of the expression of three proteins [207]. This result highlights that proteomics and the study of expression rates of particular proteins can be a powerful method in the future in order to determine if marine animals have been exposed to PAHs present in oil.

Author details

Daniela M. Pampanin
Biomiljø, International Research Institute of Stavanger, Mekjarvik, Randaberg, Norway

Magne O. Sydnes
Faculty of Science and Technology, University of Stavanger, Stavanger, Norway

Acknowledgement

Funding from the University of Stavanger is gratefully acknowledged. Andrea Bagi, University of Stavanger, is thanked for valuable literature input for the introduction part of the chapter. Emily Lyng, International Research Institute of Stavanger, is acknowledged for her careful proof reading of the manuscript.

9. References

[1] Marshall AG, Rodgers RP (2004) Petroleomics: the next grand challenge for chemical analysis. Acc. Chem. Res. 37:53-59.
[2] Fieser LF, Fieser M (1956) Organic Chemistry, 3rd edition. Boston: D. C. Heath and Co. Chapter 21.
[3] McElroy AE, Bates S, Rice SD, Korn S (1985) Bioavailability of polycyclic aromatic hydrocarbons in the aquatic environment. In: Varanasi U, editor. Metabolism of polycyclic aromatic hydrocarbons in the aquatic environment. Boca Raton: CRC Press. pp. 1-39.
[4] Feng X, Pisula W, Müllen K (2009) Large polycyclic aromatic hydrocarbons: Synthesis and discotic organization. Pure Appl. Chem. 81:2203-2224.
[5] Hylland K (2006) Polycyclic aromatic hydrocarbon (PAH) ecotoxicology in marine ecosystems. J. Toxicol. Environ. Health, Part A 69:109-123.
[6] Lang KF, Buffleb H, Kalowy J (1962) 2-Phenyl-phenanthren und binaphthyl-(2,2') aus steinkohlenteer. Chem. Ber. 95:1052-1053.
[7] Lang KF, Buffleb H, Kalowy J (1964) Fulminen (1,2-benzo-picen) im steinkohlenteer. Chem. Ber. 97:494-497.
[8] Wakeham SG, Schaffner C, Giger W (1980) Polycyclic aromatic hydrocarbons in recent lake sediments – I. Compounds having anthropogenic origins. Geochim. Cosmo. Acta. 44:403-413.

[9] Laughlin RB, Neff JM (1979) Interactive effects of salinity, temperature and polycyclic aromatic hydrocarbons on the survival and development rate of larvae of the mud crab *Rhithropanopeus harrisii*. Marine Biology 53:281-291.

[10] Harvey RG (1996) Polycyclic aromatic hydrocarbons. New York: Wiley-VCH. pp. 1-20.

[11]Achten C, Hofmann T (2010) Umweltrelevanz von natülichen polyzyklischen aromatischen kohlenwassertoffen aus steinkohlen – eine übersicht. Grundwasser 15:5-18.

[12] Durand C, Ruban V, Amblès A, Oudot J (2004) Characterization of the organic matter of sludge: Determination of lipids, hydrocarbons and PAHs from road retention/infiltration ponds in France. Environ. Pollut. 132:375-384.

[13] Beyer J, Aas E, Borgenvik HK, Ravn P (1998) Bioavailability of PAH in effluent water from an aluminium works evaluated by transplant caging and biliary fluorescence measurements of Atlantic cod (*Gadus morhua* L.). Mar. Environ. Res. 46:233-236.

[14] Næs K, Oug E (1998) The distribution and environmental relationships of polycyclic aromatic hydrocarbons (PAHs) in sediments from Norwegian smelter-affected fjords. Chemosphere 36:561-576.

[15] Smith JN, Levy EM (1990) Geochronology for polycyclic aromatic hydrocarbon contamination in sediments of the Saguenay Fjord. Environ. Sci. Technol. 24:874-879.

[16] Mascarelli A (2010) After the oil. Nature 467:22-24.

[17] Redondo J, Platonov AK (2009) Self-similar distribution of oil spills in European coastal waters. Environ. Res. Lett. 4:014008.

[18] Tedesco SA (1985) Surface geochemistry in petroleum exploration. Chapman & Hall, New York.

[19] Røe Utvik T (1999) Chemical characterization of produced water from four offshore oil production platforms in the North Sea. Chemosphere 39:2593-2606.

[20] Hunt JM (1979) Petroleum geochemistry and geology. San Francisco: W. H. Freeman and Co. p 617.

[21] Bakke T, Hameedi J, Kimstach V, Macdonald R, Melnikov S, Robertson A, Shearer R, Thomas D (1998) Petroleum hydrocarbons. In: Roberts A, editor. AMAP Assessment Report: Arctic pollution issues. pp 667-668.

[22] van der Meer F, van Dijk P, van der Werff H, Yang H (2002) Remote sensing and petroleum seepage: a review and case study. Terra Nova 14:1-17.

[23] NRC (National Research Council) (1985) Oil in the sea: Inputs, fates, and effects. Washington: National Academy Press. p. 501.

[24] Kemsley J (2012) Water eased oil removal in Gulf. Chemical and Engineering News, February 6:32-33.

[25] Head IM, Jones DM, Röling WFM (2006) Marine microorganisms make a meal of oil. Nat. Rev. Microbiol. 4:173-182.

[26] Valentine DL, Mezić I, Maćešić S, Crnjarić-Žic N, Ivić S, Hogan PJ, Fonoberov VA, Loire S (2012) Dynamic autoinoculation and the microbial ecology of a deep water hydrocarbon irruption. Proc. Natl. Acad. Sci. USA, doi: 10.1073/pnas.1108820109.

[27] Redmond MC, Valentine DL (2012) Natural gas and temperature structured a microbial community response to the Deepwater Horizon oil spill. Proc. Natl. Acad. Sci. USA, doi: 10.1073/pnas.1108756108.

[28] Atlas RM (2011) Oil biodegradation and bioremediation: A tale of the two worst spills in U.S. history. Environ. Sci. Technol. 45:6709-6715.

[29] Pothuluri JV, Cerniglia CE (1994) Microbial metabolism of polycyclic aromatic hydrocarbons. In: Chaudhry GR, editor. Biological degradation and bioremediation of toxic chemicals. London: Chapman & Hall. pp. 92-124.

[30] Volkering F, Breure, AM, Sterkenburg A, van Andel JG (1992) Microbial degradation of polycyclic aromatic hydrocarbons: effect of substrate availability on bacterial growth kinetics. Appl. Microbiol. Biotechnol. 36:548-552.

[31] Geiselbrecht AD, Herwig RP, Deming JW, Staley JT (1996) Enumeration and phylogenetic analysis of polycyclic aromatic hydrocarbon-degrading marine bacteria from Puget Sound sediments. Appl. Environ. Microbiol. 62:3344-3349.

[32] Geiselbrecht AD, Hedlund BP, Tichi MA, Staley JT (1998) Isolation of marine polycyclic aromatic hydrocarbon (PAH)-degrading *Cycloclasticus* strains from the Gulf of Mexico and comparison of their PAH degradation ability with that of Puget Sound. Appl. Environ. Microbiol. 64:4703-4710.

[33] Cao B, Nagarajan K, Loh K-C (2009) Biodegradation of aromatic compounds: current status and opportunities for biomolecular approaches. Appl. Microbiol. Biotechnol. 85:207-228.

[34] Haritash AK, Kaushik CP (2009) Biodegradation aspects of polycyclic aromatic hydrocarbons (PAHs): A review. J. Hazard. Mat. 169:1-15

[35] Keith LH, Telliard WA (1979) Priority pollutants: I. A perspective view. Environ. Sci. Technol. 13:416-423.

[36] Kerr JM, Melton HR, McMillen SJ, Magaw RI, Naughton G, Little GN (1999) Polyaromatic hydrocarbon content in crude oils around the world. Conference paper from the 1999 SPE/EPA Exploration and production environmental conference held in Austin, Texas, USA, 28 February-3 March.

[37] Faksness L-G, Brandvik PJ, Sydnes LK (2008) Composition of the water accommodated fractions as a function of exposure times and temperatures. Mar. Pollut. Bull. 56:1746-1754.

[38] Aas E, Baussant T, Balk L, Liewenborg B, Andersen OK (2000) PAH metabolites in bile, cytchrome P4501A and DNA adducts as environmental risk parameters for chronic oil exposure: a laboratory experiment with Atlantic cod. Aquat. Toxicol. 51:241-258.

[39] Deepthike HU, Tecon R, van Kooten G, von der Meer JR, Harms H, Wells M, Short J (2009) Unlike PAHs from Exxon Valdez crude oil, PAHs from Gulf of Alaska coals are not readily bioavailable. Environ. Sci. Technol. 43:5864-5870.

[40] Sundt RC, Beyer J, Vingen S, Sydnes MO (2011) High matrix interference affecting detection of PAH metabolites in bile of Atlantic hagfish (*Myxine glutinosa*) used for biomonitoring of deep-water oil production. Mar. Environ. Res. 71:369-374.

[41] Rhodes S, Farwell A, Hewitt LM, MacKinnon M, Dixon DG (2005) The effects of dimethylated and alkylated polycyclic aromatic hydrocarbons on the embryonic development of the Japanese medaka. Ecotox. Environ. Safety 60:247-258.

[42] Carls MG, Holland L, Larsen M, Collier TK, Scholz NL, Incardona JP (2008) Fish embryos are damaged by dissolved PAHs, not oil particles. Aquat. Toxicol 88:121-127.

[43] Billiard SM, Querbach K, Hodson PV (1999) toxicity of retene to early life stages of two freshwater fish species. Environ. Toxicol. Chem. 18:2070-2077.

[44] White KL (1986) An overview of immunotoxicology and carcinogenic polycyclic aromatic hydrocarbons. J Environ. Sci. Health Part C: Environ. Carcino. Ecotox. Rev. 2:163-202.

[45] Conney AH (1982) Induction of microsomal enzymes by foreign chemicals and carcinogenesis by polycyclic aromatic hydrocarbons. Cancer Res. 42:4875-4917.

[46] Canestro D, Raimondi PT, Reed DC, Schrnitt RJ, Holbrook SJ (1996) A study of methods and techniques for detecting ecological impacts. American Academy of Underwater Sciences (AAUS).

[47] Zhu SQ, King SC, Haasch ML (2008) Biomarker induction in tropical fish species on the Northwest Shelf of Australia by produced formation water. Mar. Environ. Res. 65:315-324.

[48] Hylland K, Tollefsen KE, Ruus A, Jonsson G, Sundt RC, Sanni S, Røe Utvik TI, Johnsen S, Nilssen I, Pinturier L, Balk L, Baršienė J, Marigòmez I, Feist SW, Børseth JF (2008) Water column monitoring near oil installations in the North Sea 2001–2004. Mar. Poll. Bull. 56:414–429.

[49] Gorbi S, Virno Lamberti C, Notti A, Benedetti M, Fattorini D, Moltedo G, Regoli F (2008) An ecotoxicological protocol with caged mussels, Mytilus galloprovincialis, for monitoring the impact of an offshore platform in the Adriatic Sea. Mar. Environ. Res. 65:34-49.

[50] Crone TJ, Tolstoy M (2010) Magnitude of the 2010 Gulf of Mexico oil leak. Science 330:634.

[51] Peterson CH, Rice SD, Short JW, Esler D, Bodkin JL, Ballachey BE, Irons DB (2003) Long-term ecosystem response to the Exxon Valdez oil spill. Science 302:2082-2086.

[52] Short J, Rice SD, Heintz RA, Carls MG, Moles A (2003) Long-term effects of crude oil on developing fish: Lessons from the Exxon Valdez oil spill. Energy Sources 25:509-517.

[53] Lavrova OY, Kostianoy AG (2011) Catastrophic oil spill in the Gulf of Mexico in April-May 2010. Atmospheric and Oceanic Physics 47:1114-1118.

[54] Sundt RC, Pampanin DM, Grung M, Baršienė, Ruus A (2011) PAH body burden and biomarker responses in mussel (Mytilus edulis) exposed to produced water from a

North Sea oil field: Laboratory and field assessments. Mar. Poll. Bull. 62:1498-1505

[55] Sundt RC, Ruus A, Jonsson H, Skarphéðinsdóttir H, Meier S, Grung M, Beyer J, Pampanin DM (2012) Biomarker responses in Atlantic cod (*Gadus morhua*) exposed to produced water from a North Sea oil field: Laboratory and field assessments Original Research Article. Mar. Poll. Bull. 64:144-152.

[56] Røe TI, Johnsen S (1996) Discharges of produced water to the North Sea; Effects in the water column. Produced water 2. Environmental Issues and Mitigation Technologies. S. Johnsen. New York: Plenum Press pp. 13-25.

[57] Durell G, Utvik TR, Johnsen S, Frost T, Neff J (2006) Oil well produced water discharges to the North Sea. Part I: Comparison of deployed mussels (Mytilus edulis), semi-permeable membrane devices, and the DREAM model predictions to estimate the dispersion of polycyclic aromatic hydrocarbons. Mar. Environ. Res. 62:194-223.

[58] OGP (2005) Fate and effects of naturally occurring substances in produced water on the marine environment. International Association of Oil & Gas Producers, London, UK, Report No 364. 36 p.

[59] Johnsen S, Røe Utvik TI, Garland E, de Vals B, Campbell J (2004) Environmental fate and effects of contaminants in produced water. SPE 86708. Paper presented at the Seventh SPE international conference on health, safety and environment in oil and gas exploration and production. Society of Petroleum Engineers, Richardson, TX, 9 p.

[60] Latimer JS, Zheng J (2003) The sources, transport and fate of PAHs in the marine environment. In: Doube PET editor. PAHs: an ecotoxicological perspective. Wiley, West Sussex, pp. 9-34.

[61] Neff J, Lee K, De Blois EM (2011) Produced water: overview of composition, fate, and effects. In: Lee K, Neff J editors. Produced Water. Environmental risks and advances in mitigation technologies. Springer, London, UK. pp 3-54.

[62] Ruus A, Tollefsen KE, Grung M, Klungsøyr J, Hylland K (2006) Accumulation of contaminations in pelagic organisms, caged blue mussels, caged cod and semi-permeable membrane device (SPMDs). In: Hylland K, Vethaak AD, Lang T, editors. Biological effects of contaminants in marine pelagic ecosystems (ICES). SETAC publications pp 51-74.

[63] Faksness LG, Grini PG, Daling PS (2004) Partitioning of semi-soluble organic compounds between the water phase and oil droplets in produced water. Mar. Poll. Bull. 48:731-742.

[64] Burns KA, Codi S (1999) Non-volatile hydrocarbon chemistry studies around a production platform on Australia's northwest shelf. Estur. Cstl. Shelf Sci. 49:853-876.

[65] Neff JM, Johnsen S, Frost TK, Røe Utvik TI, Durell GS (2006) Oil well produced water discharges to the North Sea. Part II: comparison of deployed mussels (*Mytilus edulis*) and the DREAM model to predict risk assessment. Mar. Environ. Res. 62:224-246.

[66] OLF (2007) Environmental report 2007. The Norwegian Oil Industry Association (OLF). Stavanger, Norway 63 p.
[67] Farmen E, Harman C, Hylland K, Tollefsen KE (2010). Produced water extracts from North Sea oil production platforms result in cellular oxidative stress in a rainbow trout in vitro bioassay. Mar. Poll. Bull. 60:1092-1098.
[68] Suter GW (1993) Ecological Risk Assessment. Lewis Publishers, Boca Raton, FL, USA 538 p.
[69] Henderson F, Bechtold WE, Bond JA, Sun JD (1989) The use of biological markers in toxicology. Crit. Rev. Toxicol. 20:65–82.
[70] De Zwart D (1995) Monitoring water quality in the future, Volume 3: Biomonitoring. National Institute of Public Health and Environmental Protection (RIVM), Bilthoven, The Netherlands, 81 p.
[71] Antizar-Ladislao B (2009) Polycyclic aromatic hydrocarbons, polycholirnated biphenyls, phthalates and organotins in northern Atlantic Spain's coastal marine sediments. J. Environ. Monitor. 11:85-91.
[72] Binelli A, Provini A (2003) POPs in edible clams from different Italian and European markets and possible human health risk. Mar. Poll. Bull. 46:879-886.
[73] Water Framework Directive (2000) Directive 2000/60 EU of the European Parliament and of the Council of 23 October 2000 establishing a framework for Community action in the field of water policy.
[74] OSPAR Commission (2000) Quality Status Report 2000. London: OSPAR.
[75] Gilroy (2000) Derivation of shellfish harvest reopening criteria following the new Carissa oil spill in Coos Bay, Oregon. J. Toxicol. Environ. Health 60:317-329.
[76] Senthil Kumar K, Sajwan KS, Richardson JP, Kannan K (2008) Contamination profiles of heavy metals, organochlorine pesticides, polycyclic aromatic hydrocarbons and alkylphenols in sediment and oyster collected from marsh/estuarine Savannah GA, USA. Mar. Poll. Bull. 56:136-162.
[77] Massara Paletto V, Commendatore MG, Esteves JL (2008) Hydrocarbon levels in sediments and bivalve mollusks from Bahía Nueva (Patagonia, Argentina): an assessment of probable origin and bioaccumulation factors. Mar. Poll. Bull. 56:2082-2105.
[78] Francioni EL, Wagener A de LR, Scofield A, Depledge M, Sette CB, Carvalhosa L (2007) Polycyclic aromatic hydrocarbon in mussel *Perna perna* from Guanabara Bay, Brazil: space-time observations, source investigation and genotoxicity. Sci. Tot. Environ. 372:515-531.
[79] Baumard P, Budzinski H, Garrigues P, Narbonne JF, Burgeot T, Miche X et al. (1999) Polycyclic aromatic hydrocarbon (PAH) burden of mussels (*Mytilus* sp.) in different marine environments in relation with sediment PAH contamination and bioavailability. Mar. Environ. Res. 47:415-439.
[80] David A, Fenet H, Gomez E (2009) Alkylphenols in marine environments: distribution monitoring strategies and detection considerations. Mar. Poll. Bull. 58:953-960.
[81] Barbee GC, Duncan JBB, Bickham JW, Matson CW, Hintze CJ, Autenrieth RL, ZhouG-D, McDonald TJ, Cizmas L, Norton D, Donnelly KC (2008) *In situ* biomonitoring of PAH-

contaminated sediments using juvenile coho salmon (*Oncorhynchus kisutch*). Ecotoxicol. Environ. Safety 71:454-464.

[82] HMSO (1985) The determination of 6 specific PAHs. Materials for examination of waters and associated materials. Her Majesty's Stationery Office, London.

[83] EPA Procedure 8310 (1986) Polycyclic aromatic hydrocarbons.

[84] EPA Method 8270D (1998) (SW-846) Semivolatile organic compounds by gas chromatography/mass spectrometry (GC/MS), Revision 4.

[85] ISO Procedure 1799 (2002) Water quality-determination of 15 PAHs in water by HPLC with fluorescence detection.

[86] Wolska L (2008) Determination (monitoring) of PAHs in surface waters: why an operationally defined procedure is needed. Anal. Bioanal. Chem. 391:2647-2652.

[87] Beyer J, Jonsson G, Porte C, Krahn MM, Ariese F (2010) Analytical methods for determining metabolites of polycyclic aromatic hydrocarbon (PAH) pollutants in fish bile: a review. Environ. Toxicol. Pharmacol. 30:224-244.

[88] Neff JM (2002) Bioaccumulation in marine organisms. Effect of contaminants from oil well produced water. Oxford, Elsevier Science Ltd., 468 p.

[89] Ardgar H, Ozel MZ, Sen Z (2011) Polycyclic aromatic hydrocarbons in water from the Menderes river, turkey. Bull. Environ. Contam. Toxicol. 86:221-225.

[90] Oros DR, Ross JRM, Spies RB, Mumley T (2007) Polycyclic aromatic hydrocarbon (PAH) contamination in San Francisco Bay: a 10-year retrospective of monitoring in an urbanized estuary. Environ. Res. 105:101-118.

[91] Cataldo D, Colombo JC, Boltovskoy D, Bilos C, Landoni P (2001) Environmental toxicity assessment in the Paraná river delta (Argentina): simultaneous evaluation od selected pollutants and mortality rates of *Curbilcula fluminea* (bivalvia) early juveniles. Environ. Poll. 112:379-389.

[92] Colombo JC, Barreda C, Bilos NC, migota MC, Skorupka C (2005) Oil spill in the Rio de la Plata estuary, Argentina: 2-hydrocarbon disappearance rates in sediments and soils. Environ. Poll. 134:267-276.

[93] Arias AH, Spetter CV, Freije RH, Marcovecchio JE (2009) Polycyclic aromatic hydrocarbons in water, mussel (*Brachidontes* sp., *Tagelus* sp.) and fish (*Odontesthes* sp.) from Bahía Blanca Estuary, Argentina. Estuar. Coastal Shelf Sci. 85:67-81.

[94] Allan SE, Smith BW, Anderson KA (2012) Impact of the Deepwater Horizon Oil Spill on Bioavailable Polycyclic Aromatic Hydrocarbons in Gulf of Mexico Coastal Waters. Environ. Sci. Technol. 46:2033-2039.

[95] Viarengo A, Dondero F, Pampanin DM, Fabbri R, Poggi E, Malizia M, Bolognesi C, Perrone E, Gollo E, Cossa GP (2007) A biomonitoring study assessing the residual biological effects of pollution caused by the HAVEN wreck on marine organisms in the Ligurian sea (Italy). Arch. Environ. Contam. Toxicol. 53:607-616.

[96] Martinez-Gomez C, Fernandez B, Valdes J, Campillo JA, Benedicto J, Sanchez F, Vethaak AD (2009) Evaluation of three-year monitoring with biomarkers in fish following the Prestige oil spill (N Spain). Chemosphere 74:613-620.

[97] Uno S, Koyama J, Kokushi E, Monteclaro H, Santander S, Cheikyula JO, Miki S, Anasco N, Pahila IG, Taberna Jr HS, Matsuoka T (2010) Monitoring of PAHs and alkylated PAHs in aquatic organisms after 1 month from the Solar I oil spill off the coast of Guimaras Island, Philippines. Environ. Monit. Assess. 165:501-515.

[98] Lee K, Neff J (2011) Produced Water. Environmental risks and advances in mitigation technologies. Springer, New York, 608 p.

[99] Harman C, Thomas K, Tollefsen KE, Meier S, Bøyum O, Grung M (2009) Monitoring the freely dissolved concentrations of polycyclic aromatic hydrocarbons (PAH) and alkylphenols (AP) around a Norwegian oil platform by holistic passive sampling. Mar. Poll. Bull. 58:1671-1679.

[100] Harman C, Tollefsen KE, Bøyum O, Thomas K, Grung M (2008) Uptake rates of alkylphenols, PAHs and carbazoles in semipermeable membrane devices (SPMDs) and polar organic chemical integrative samplers (POCIS). Chemosphere 72:1510-1516.

[101] Sundt RC, Brooks S, Grøsvik BE, Pampanin DM, Farmen E, Harman C, Meier S (2010) Water column monitoring of offshore produced water discharges. Compilation of previous experience and suggestions for future survey design. OLF 2010.

[102] Livingstone DR (1998) The fate of organic xenobiotics in aquatic ecosystems: quantitative and qualitative differences in biotransformation by invertebrates and fish. Comp. Biochem. Physiol. A 120:43-49.

[103] Fillmann G, Watson GM, Howsam M, Francioni E, Depledge MH, Readman JW (2004) Urinary PAH metabolites as biomarkers of exposure in aquatic environments. Environ. Sci. Technol. 38:2649-2656.

[104] van der Oost R, Beyer J, Vermeulen NPE (2003) Fish bioaccumulation and biomarkers in environmental risk assessment: a review. Environ. Toxicol. Pharmacol. 13:57-149.

[105] Bellas J, Saco-Álvarez L, Nieto Ó, Beiras R (2008) Ecotoxicological evaluation of polycyclic aromatic hydrocarbons using marine invertebrate embryo–larval bioassays. Mar. Poll. Bull. 57:493-502.

[106] Aas E, Beyer J, Jonsson G, Reichert WL, Andersen OK (2001) Evidence of uptake, biotransformation and DNA binding of polycyclic aromatic hydrocarbons in Atlantic cod and corkwing wrasse caught in the vicinity of an aluminium works. Mar. Environ. Res. 52:213-229.

[107] Oros DR, Ross JRM (2005) Polycyclic aromatic hydrocarbons in bivalves from the San Francisco estuary: Spatial distributions, temporal trends, and sources (1993-2001). Mar. Environ. Res. 60:466-488.

[108] Sole M, Buet A, Ortiz L, Maynou F, Bayona JM, Albaiges J (2007) Bioaccumulation and biochemical responses in mussels exposed to the water-accommodated fraction of the Prestige fuel oil. Sci. Mar. 71:373-382.

[109] Røe Utvik TI, Durell GS, Johnsen S (1999) Determining produced water originating polycyclic aromatic hydrocarbons n North Sea waters: comparison of sampling techniques. Mar. Poll. Bull. 38:977-989.

[110] Grøsvik BE, Meier S, Liewenborg B et al. (2009) Condition monitoring in the water column 2008: Oil hydrocarbons in fish from Norwegian waters. IMR report Nr.2-2009.

[111] Hylland K, Lang T, Vethaak D, editors (2006) Biological Effects of Contaminants in Marine Pelagic Ecosystems. SETAC Press, 475 p.

[112] Shaw GR, Connell DW (2001) DNA adducts as a biomarker of polycyclic aromatic hydrocarbon exposure in aquatic organisms: relationship to carcinogenicity. Biomarkers 6:64-71.

[113] Kammann U, Lang T, Vobach M, Wosniok W (2005) Ethoxyresorufin-O-deethylase (EROD) activity in dab (Limanda limanda) as biomarker for marine monitoring. Environ. Sci. Pollut. Res. 12:140-145.

[114] Balk L, Hylland K, Hansson T, Bertssen MHG, Beyer J, Jonsson G, Melbey A, Grung M, Torstensen BE, Børset JF, Skarphéðinsdóttir H, Klungsøyr J (2011) Biomarkers in natural fish populations indicate adverse biological effects of offshore oil production. PloS ONE 6:1-10.

[115] Maria VL, Correia AC, Santos MA (2002) Benzo[a]pyrene and beta-naphthoflavone mutagenic activation by European eel (*Anguilla anguilla* L.) S9 liver fraction. Ecotoxicol. Environ. Saf. 53:81-85.

[116] Stephensen E, Adolfsson-Erici M, Celander M, Hulander M, Parkkonen J, Hegelund T, Sturve J, Hasselberg L, Bengtsson M, Forlin L (2003) Biomarker responses and chemical analyses in fish indicate leakage of polycyclic aromatic hydrocarbons and other compounds from car tire rubber. Environ. Toxicol. Chem. 22:2926-2931.

[117] Pathiratne A, Hemachandra CK (2010) Modulation of ethoxyresorufin O-deethylase and glutathione S-transferase activities in Nile tilapia (*Oreochromis niloticus*) by polycyclic aromatic hydrocarbons containing two to four rings: implications in biomonitoring aquatic pollution. Ecotoxicol. 19:1012–1018.

[118] Bravo, CF, Curtis LR, Myers MS, Meador JP, Johnson LL, Buzitis J, Collier TK, Morrow JD, Laetz CA, Loge FJ, Arkoosh MR (2011) Bioamrker responses and disease susceptibility in juvenile rainbow trout *Oncorhynchus mykiss* fed a high molecular weight PAH mixture. Environ. Toxicol. Chem. 30:704-714.

[119] Broeg K, Zander S, Diamant A, Korting W, Kruner G, Paperna I, von Westernhagen H (1999) The use of fish metabolic, pathological and parasitological indices in pollution monitoring - 1. North Sea. Helgoland Mar. Res. 53:171-194.

[120] Cajaraville MP, Bebianno MJ, Blasco J, Porte C, Sarasquete C, Viarengo A (2000) The use of biomarkers to assess the impact of pollution in coastal environments of the Iberian Peninsula: a practical approach. Sci. Tot. Environ. 247:295-311.

[121] Pluta HJ (1993) Investigations on biotransformation (mixed function oxygenase activities) in fish liver. In: Braunbeck T, Hanke W, Segner H editors. Fish ecotoxicology and ecophysiology. VCH Weinheim pp. 13-33.

[122] Stegeman JJ , Hahn ME (1994) Biochemistry and molecular biology of monooxygenases: current perspectives on forms, functions, and regulation of cytochrome P450 in aquatic species. In: Malins DC, Ostrander GK, editors. Boca Raton: Aquatic Publishers pp. 87–203.

[123] Jőnsson EM, Brandt I, Brunstrom B (2002) Gill filament-based EROD assay for monitoring waterborne dioxin-like pollutants in fish. Environ. Sci. Technol. 36:3340-3344.

[124] Dissanayake A, Bamber SD (2010) Monitoring PAH contamination in the field (South west Iberian Peninsula): biomonitoring using fluorescence spectrophotometry and physiological assessment in the shore crab *Carcinus maenas* (L.) (Crustacea: Decapoda). Mar. Environ. Res. 70:65-72.

[125] Boyd DR, Kennedy DA, Malone JF, O'Kane GA (1987) Synthesis of triphenylene 1,2-oxide (1,2-epoxy-1,2-dihydrophenylene) and absolute configuration of the *trans*-1,2-dihydro diol metabolite of triphenylene. Crystal structure of (-)-(1*R*,2*R*)-*trans*-2-bromo-1-menthyloxyacetoxy-1,2,3,4-tetrahydrotiphenylene. J. Chem. Soc. Perkin Trans. 1 369-375.

[126] Rand GM (1995) Fundamentals of aquatic toxicology, second edition: Effects, environmental fate and risk assessment. Washington DC: Taylor and Francis. 128 p.

[127] Nebert DW, Dalton TP, Okey AB, Gonzalez FJ (2004) Role of aryl hydrocarbon receptor-mediated induction of the CYP1 enzymes in environmental toxicity and cancer. J. Biol. Chem. 279:23847-23850.

[128] Jacob J (2008) The significance of polycyclic aromatic hydrocarbons as environmental carcinogens. 35 Years research on PAH-a retrospective. Polycycl. Aromat. Comp. 28:242-272.

[129] Uno T, Ishizuka M, Itakura T (2012) Cytochrome P450 (CYP) in fish. Environ. Toxicol. Pharmacol. 34:1-13.

[130] Jacob J, Raab G, Soballa V, Schmalix WA, Grimmer G, Greim H, Doehmer J, Seidel A (1996) Cytochrom P450-mediated activation of phenanthrene in genetically engineered V79 Chinese hamster cells. Environ. Toxicol. Pharmacol. 1:1-11.

[131] Pangrekar J, Kandaswami C, Kole P, Kumar S, Sikka HC (1995) Comparative metabolism of benzo(*a*)pyrene, chrysene and phenanthrene by brown bullhead liver microsomes. Mar. Environ. Res. 39:51-55.

[132] Pangrekar J, Kole PL, Honey SA, Kumar S, Sikka HC (2003) Metabolism of chrysene by brown bullhead liver microsomes. Toxicol. Sci. 71:67-73.

[133] Shou M, Korzekwa KR, Krausz KW, Crespi CL, Gonzales FJ, Gelboin HV (1994) Regio- and stereo-selective metabolism of phenanthrene by twelve cDNA-expressed human, rodent, and rabbit cytochrome P-450. Cancer Lett. 83:305-313.

[134] Jacob J, Doehmer J, Grimmer G, Soballa V, Raab G, Seidel A, Greim H (1996) Metabolism of phenanthrene, benz[*a*]anthracene, benzo[*a*]pyrene, chrysene and

benzo[*c*]phenanthrene by eight cDNA-expressed human and rat cytochromes P450. Polycycl. Aromat. Comp. 10:1-9.

[135] Jacob J, Raab G, Soballa VJ, Luch A, Grimmer G, Greim H, Doehmer J, Morrison HL, Stegeman JJ, Seidel A (1999) Species-dependent metabolism and benzo[*c*]phenanthrene and dibenzo[*a,l*]pyrene by various CYP450 isoforms. Polycycl. Aromat. Comp. 16:191-203.

[136] Seidel A, Soballa VJ, Raab G, Frank H, Greim H, Grimmer G, Jacob J, Doehmer J (1998) Regio- and stereoselectivity in the metabolism of benzo[*c*]phenanthrene mediated by genetically engineered V79 Chinese hamster cells expressing rat and human cytochromes P450. Environ. Toxicol. Pharmacol. 5:179-196.

[137] Schober W, Luch A, Soballa VJ, Raab G, Stegeman JJ, Doehmer J, Jacob J, Seidel A (2006) On the species-specific biotransformation of debenzo[*a,l*]pyrene. Chemico.-Biol. Interact 161:37-48.

[138] Islam NB, Whalen DL, Yagi H, Jerina DM (1987) pH Dependence of the mechanism of hydrolysis of benzo[*a*]pyrene-*cis*-7,8-diol 9,10-epoxide catalyzed by DNA, poly(G), and poly(A). J. Am. Chem. Soc. 109:2108-2111.

[139] Thakker DR, Yagi H, Levin W, Wood AW, Conney AH, Jerina DM (1985) Polycyclic aromatic hydrocarbons: Metabolic activation to ultimate carcinogens. In: Ander MW, editor. Bioactivation of Foreign Compounds. Orlando: Academic Press. pp. 177-242.

[140] Jonsson G, Taban IC, Jørgensen KB, Sundt RC (2004) Quantitative determination of de-conjugated chrysene metabolites in fish bile by HPLC-fluorescence and GC-MS. Chemosphere 54:1085-1097.

[141] Wenzel-Hartung R, Brune H, Grimmer G, Germann P, Timm J, Wosniok W (1990) Evaluation of the carcinogenic potency of 4 environmental polycyclic aromatic compounds following intrapulmonary application in rats. Exp. Pathol. 40:221-227.

[142] Harvey RG (1991) Polycyclic aromatic hydrocarbons: chemistry and carcinogenesis. Cambridge: Cambridge University Press. pp. 26-49.

[143] Glatt H, Wameling C, Elsberg S, Thomas H, Marquardt H, Hewer A, Phillips DH, Oesch F, Seidel A (1993) Genotoxicity characteristics of reverse diol-epoxides of chrysene. Carcinogenesis 14:11-19.

[144] Grimmer G, Brune H, Dettbarn G, Heinrich U, Jacob J, Mohtashamipur E, Norpoth K, Pott F, Wenzelhartung R (1988) Urinary and fecal excretion of chrysene and chrysene metabolites by rats after oral, intraperitoneal intratracheal or intrapulmonary application. Arch. Toxicol. 62:401-405.

[145] Bae S, Mah H, Chaturvedi S, Jeknic TM, Baird WM, Katz AK, Carrell HL, Glusker JP, Okazaki T, Laali KK, Zajc B, Lakshman MK (2007) Synthetic, crystallographic, computational, and biological studies of 1,4-difluorobenzo[*c*]phenanthrene and its metabolites. J. Org. Chem. 72:7625-7633.

[146] Buening MK, Wislocki PG, Levin W, Yagi H, Thakker DR, Akagi H, Koreeda M, Jerina DM, Conney AH (1978) Tumorigenicity of the optical enantiomers of the diastereomeric benzo[*a*]pyrene 7,8-diol-9,10-epoxides in newborn mice: Exceptional activity of (+)-7beta,8alpha-dihydroxy-9alpha,10alpha-epoxy-7,8,9,10-tetrahydrobenzo[*a*]pyrene. Proc. Natl. Acad. Sci. USA 75:5358-5361.

[147] Slaga TJ, Bracken WB, Gleason C, Levin W, Yagi H, Jerina DM, Conney AH (1979) Marked differences in the skin tumor-initiating activities of the optical enantiomers of the diastereomeric benzo(*a*)pyrene 7,8-diol-9,10-epoxides. Cancer Res. 39:67-71.

[148] Luch A, Seidel A, Glatt H, Platt KL (1997) Metabolic activation of the (+)-*S*,*S*- and (-)-*R*,*R*-enantiomers of *trans*-11,12-dihydroxy-11,12-dihydrodibenzo[*a*,*l*]pyrene: Stereoselectivity, DNA adduct formation, and mutagenicity in Chinese hamster V79 cells. Chem. Res. Toxicol. 10:1161-1170.

[149] Miller EC (1978) Some current perspectives on chemical carcinogenesis in humans and experimental animals: Presidential address. Cancer Res. 38:1479-1496.

[150] Jarina DM, Lehr RE, Yagi H, Hermandez O, Dansette PM, Wislocki PG, Wood AW, Chang RL, Levin W, Conney AH (1976) Mutagenicity of benzo[*a*]pyrene derivatives and the description of a quantum mechanical model which predicts the ease of carbonium ion formation from diol epoxides. In: de Serres FJ, Fouts JR, Bend JR, Philpot RM, editors. *In vitro* metabolic activation in mutagenesis testing. Amsterdam: Elsevier. pp 159-177.

[151] Dipple A, Moschel RC, Bigger CAH (1984) Polynuclear aromatic carcinogens. In: Searle CE, editor. *Chemical carcinogens*, 2nd edition. ACS monograph 182; Vol. 1. Washington DC: American Chemical Society. pp 41-163.

[152] Grover PL (1979) Chemical carcinogens and DNA; Vol 1. Boca Raton: CRC Press. 236 p.

[153] Grover PL (1979) Chemical carcinogens and DNA; Vol 2. Boca Raton: CRC Press. 210 p.

[154] Neidle S (1980) Carcinogens and DNA. Nature 283:135.

[155] Miller JA (1970) Carcinogenesis by chemicals – an overview (G. H. A. Clowes Memorial Lecture). Cancer Res. 30:559-576.

[156] Szeliga J, Dipple A (1998) DNA adduct formation by polycyclic aromatic hydrocarbon dihydrodiol epoxides. Chem. Res. Toxicol. 11:1-11.

[157] Melendez-Colon VJ, Luch A, Seidel A, Baird WM (1999) cancer initiation by polycyclic aromatic hydrocarbons results from formation of stable DNA adducts rather than apurinic sites. Carcinogenesis 20:1885-1891.

[158] Stein JE, Reichert WL, French B, Varanasi U (1993) ^{32}P-Postlabeling analysis of DNA adduct formation and persistence in English sole (*Pleuroectes vetulus*) exposed to benso[*a*]pyrene and 7H-dibenzo[*c*,*g*]carbazole. Chem. –Biol. Interact. 88:55-69.

[159] van der Oost R, Heida H, Satumalay K, van Schooten FJ, Ariese F, Vermeulen NPE (1994) Bioaccumulation, biotransformation and DNA binding of pahs in feral eel (*Anguilla anguilla*) exposed to polluted sediments: A field survey. Environ. Toxicol. Chem. 13:859-879.

[160] Krahn MM, Rhodes LD, Myers MS, Moore LK, MacLeod WD, Malins DC (1986) Associations between metabolites of aromatic compounds in bile and the accurrence of hepatic lesions in English sole (*Pleuroectes vetulus*) from Puget Sound, Washington. Arch. Environ. Contam. Toxicol. 15:61-67.

[161] Dunn BP, Black JJ, Maccubbin A (1987) ^{32}P-Postlabeling analysis of aromatic DNA adducts in fish from polluted areas. Cancer Res. 47, 6543-6548.

[162] Stagg RM (1998) the development of an international programme for monitoring the biological effects of contaminants in the OSPAR convention area. Mar. Environ. Res. 46:307-313.

[163] Meyers MS, Johnson LL, Hom T, Collier TK, Stein JE, Varanasi U (1998) Toxiocopathic hepatic lesions in subadult English sole (Pleuronectes vetuls) from Puget Sound, Washington, USA: Relationships with other biomarkers of contaminant exposure. Mar. Environ. Res. 45:47-67.

[164] Cook JW, Hewett CL, Hieger I (1933) The isolation of a cancer producing hydrocarbon from coal tar. Part I, II and III. J. Chem. Soc. 395-405.

[165] Xue W, Warshawsky D (2005) Metabolic activation of polycyclic and heterocyclic aromatic hydrocarbons and DNA damage: a review. Toxicol. Appl. Pharmacol. 206:73-93.

[166] La DK, Swenberg JA (1996) DNA adducts: biological markers of exposure and potential applications to risk assessment. Mutat. Res. 365:129-146.

[167] Hsu GW, Huang X, Luneva N, Geacintov NE, Beese LS (2005) Structure of a high fidelity DNA polymerase bound to a benzo[a]pyrene adduct that blocks replication. J. Biol. Chem. 280:3764-3770.

[168] Christian TD, Romano LJ (2009) Monitoring the conformation of benzo[a]pyrene adducts in the polymerase active site using fluorescence resonance energy transfer. Biochem. 48:5382-5388.

[169] Noaksson E, Tjärnlund U, Ericson G, Balk L (1998) Biological effects on viviparous blenny exposed to chrysene and held in synthetic as well as in natural brackish water. Mar. Environ. Res. 46:81-85.

[170] Skipper PL, Tannenbaum SR (1990) Protein adducts in the molecular dosimetry of chemical carcinogens. Carcinogenesis 11:507-518.

[171] Lyons BP, Harvey JS, Parry JM (1997) An initial assessment of the genotoxic impact of the Sea Empress oil spill by the measurement of DNA adduct levels in the intertidal teleost *Lipophorys pholis*. Mutat. Res. 263-268.

[172] Harvey JS, Lyons BP, Page TS, Stewart C, Parry JM (1999) An assessment of the genotoxicity impact of the Sea Empress oil spill by measurement of DNA adduct levels in selected invertebrate and vertebrate species. Mutat. Res. 103-114.

[173] Amat A, Burgeot T, Castegnaro M, Pfohl-Leszkowicz A (2006) DNA adducts in fish following an oil spill exposure. Environ. Chem. Lett. 4:93-99.

[174] French B, Reichert WL, Hom HR, Nishimoto HR, Stein JE (1996) Accumulation and dose response of hepatic DNA adducts in English sole (*Pleuronectes vetulus*) exposed to a gradient of contaminated sediments. Aquat. Toxicol. 36:1-16.

[175] Ericson G, Noaksson E, Balk L (1999) DNA adduct formation and persistence in liver and extrahepatic tissues of northern pike (*Exos Lucius*) following oral exposure to

benzo[a]pyrene, genzo(k)fluoranthrene and 7H-dibenzo(c,g)carbazole. Mutat. Res. 427:135-145.

[176] Stein EJ, Reichert WL, Varanasi U (1994) Molecular epizootiology: assessment of exposure to genotoxic compounds in teleost. Environ. Health Perspect. 102:20-23.

[177] Varanasi U, Reichert WL, Stein JE (1989) ^{32}P-postlabelling analysis of DNA adducts in liver of wild English sole (*Parophrys vetulus*) and winter flounder (*Pseudopleuronectus americanue*). Cancer Res. 49:1171-1177.

[178] Ericson G, Larsson A (2000) DNA adduct in perch (*Perca fluviatilis*) living in coastal water polluted with bleached pulp mill effluents. Ecotox. Environ. Safety 46: 167-173.

[179] Reichert WL, Myers MS, Peck-Miller K, French B, Anulacion BF, Collier TK, Stein JE, Varanasi U (1998) Molecular epizootiology of genotoxic events in marine fish: linking contaminant exposure, DNA damage and tissues-level alterations. Mutat. Res. 411:215-225.

[180] Phillips DH (1997) Detection of DNA damage by the ^{32}P-postlabelling assay. Mutat. Res. 378:1-12.

[181] Beach AC, Gupta RC (1992) Human biomonitoring and the ^{32}P-postlabelling assay. Carcinogenesis 13:1053-1074.

[182] Reichert WL, Stein JE, French B, Goodwin P, Varanasi U (1992) Storage phosphor imaging technique for detection and quantitation of DNA adducts measured by the ^{32}P-postlabelling assay. Carcinogenesis 13:1475-1479.

[183] Randerath E, Mittal D, Randerath K (1988) Tissue distribution of covalent DNA damage in mice treated dermally with cigarette "tar". Preference for lung and heart DNA. Carcinogenesis 9:75-80.

[184] Schilderman PAEL, Moonen AJC, Maas LM, Welle I, Kleinjans JCS (1999) Use of crayfish in biomonitoring studies of environmental pollution of the river Meuse. Ecotox. Env. Saf. 44:241-252.

[185] de Almeida EA, de Almeida Marques S, Klitzke CF, Dias Bainy AC, Gennari de Medeiros MH, Di Mascio P, de Melo Moureiro AP (2003) DNa damage in digestive gland and mantle tissue of the mussel *Perna perna*. Comp. Biochem. Physiol. C 135:295-303.

[186] Wahidulla S, Rajamanickam YR (2010) Detection of DNA damage in fish *Oreochromis mossambicus* induced by co-exposure to phenanthrene and nitrite by ESI-MS/MS. Enviorn. Sci. Pollut. Res. 17:441-452.

[187] Sturla JS (2007) DNa adduct profiles: chemical approached to addressing the biological impact of DNA damage from small molecules. Curr. Opin. Chem. Biol. 11:293-299.

[188] Halliwell B (1999) Oxygen and nitrogen are procarcinogens. Damage to DNA by reactive oxygen, chlorine and nitrogen species: measurements, mechanism and the effects of nutrition. Mutat. Res. 443:37-52.

[189] Machella N, Regoli F, Cambria A, Santella RM (2004). Application od an immunoperoxidase staining method for detection of 7,8-dihydro-8-oxodeoxyguanosine

as a biomarker of chemical-induced oxidative stress in marine organisms. Aquat. Toxicol. 67:23-32.

[190] Machella N, Regoli F, Santella RM (2005) Immunofluorescent detection of 8-oxo-dG and PAH bulky adducts in fish liver and mussel digestive gland. Aquat. Toxicol. 71:335-343.

[191] Skipper PL (1996) Influence of tertiary structure on nucleophilic substitution reactions of proteins. Chem. Res. Toxicol. 9:918-923.

[192] Skipper PL (1996) Influence of tertiary structure on nucleophilic substitution reactions of proteins. Chem. Res. Toxicol. 9:918-923.

[193] Brunmark P, Harriman S, Skipper PL, Wishnok JS, Amin S, Tannenbaum SR (1997) Identification of subdomain IB in human serum albumin as a major binding site for polycyclic aromatic hydrocarbon epoxides. Chem. Res. Toxicol. 10:880-886.

[194] Day BW, Skipper PL, Zaia J, Singh K, Tannenbaum SR (1994) Enantiospecificity of covalent adduct formation by benzo[a]pyrene annti-diol epoxide with human serum albumin. Chem. Res. Toxicol. 7:829-835.

[195] Meyer MJ, Bechtold WE (1996) Protein adduct biomarkers: State of the art. Environ. Health Perspect. 104:879-882.

[196] Day BW, Doxtader MM, Rich RH, Skipper PL, Singh K, Dasari RR, Tannenbaum SR (1992) Human serum albumin-benzo[*a*]pyrene *anti*-diol epoxide adduct structure elucidation by fluorescence line narrowing spectroscopy. Chem. Res. Toxicol. 5:71-76.

[197] Day BW, Skipper PL, Zaia J, Tannenbaum SR (1991) Benzo[*a*]pyrene *anti*-diol epoxide covalently modifies human serum albumin carboxylate side chains and imidazole side chain of histidine(146). J. Am. Chem. Soc. 113:8505-8509.

[198] Zaia J, Biemann K (1994) Proteinase K activity inhibited near amino acid carrying large substituents: Three PAH diolepoxides covalently modify His-146 human serum albumin residue. J. Am. Chem. Soc. 116:7407-7408.

[199] Käfferlein HU, Marczynski B, Mensing T, Brüning T (2010) Albumin and hemoglobin adducts of benzo[*a*]pyrene in humans-Analytical methods, exposure assessment, and recommendations for future directions. Critical Rev. Toxicol. 40:126-150.

[200] Padrós J, Pelletier E (2000) In vivo formation of (+)-*anti*-benzo[*a*]pyrene diol-epoxide-plasma albumin adducts in fish. Marine Environ. Res. 50:347-351.

[201] Törnqvist M, Fred C, Haglund J, Helleberg H, Paulsson B, Rydberg P (2002) Protein adducts: quantitative and qualitative aspects of their formation, analysis and applications. J. Chromatogr. B 778:279-308.

[202] Hutchins DA, Skipper PL, Naylor S, Tannenbaum SR (1988) Isolation and characterization of major fluoranthene-hemoglobin adducts formed *in vivo* in the rat. Cancer Res. 48:4756-4761.

[203] Gromov P, Moreira JMA, Gromova I, Celis JE (2008) Proteomic strategies in bladder cancer: From tissue to fluid and back. Proteomics Clin. Appl. 2:974-988.

[204] Yim E-K, Park J-S (2006) Role of proteomics in translational research in cervical cancer. Exp. Rev. Proteomics 3:21-36.

[205] Monsinjon T, Knigge T (2007) Proteomic applications in ecotoxicology. Proteomics 7:2997-3009.

[206] Lemos MFL, Soares AMVM, Correia AC, Esteves AC (2010) Proteins in ecotoxicology – How, why and why not? Proteomics 10:873-887.

[207] Liu H, Shen J, Feng L, Yu Y (2009) Low concentration of anti-7,8-dihydroxy-9,10-epoxy-7,8,9,10-tetrahydrobenzo[*a*]pyrene induces alterations of extracellular protein profile of exposed epithelial cells. Proteomics 9:4259-4264.

Chapter 6

Phenanthrene Removal from Soil by a Strain of *Aspergillus niger* Producing Manganese Peroxidase of *Phanerochaete chrysosporium*

Diana V. Cortés-Espinosa and Ángel E. Absalón

Additional information is available at the end of the chapter

http://dx.doi.org/10.5772/51944

1. Introduction

Polycyclic aromatic hydrocarbons (PAHs) are hydrophobic compounds that have accumulated in the environment due to a variety of anthropogenic activities and their persistence is chiefly due their low water solubility. PAHs are often mutagenic and carcinogenic which emphasizes the importance of their removal from the environment [1,2]. Since the 1970s, research on the biological degradation of PAHs has demonstrated that bacteria, fungi and algae possess catabolic abilities that may be utilized for the remediation of PAH-contaminated soil and water. Phenanthrene (Phe) is one of several PAHs that are commonly found as pollutants in soils [3], estuarine waters, sediments and other terrestrial and aquatic sites [4] and has been shown to be toxic to marine diatoms, gastropods, mussels, and fish [5,6].

Solid culture systems have shown great effectiveness in the removal of toxic compounds from soil. In this method, agroindustrial wastes are used such as wheat straw, corn stalks, sugarcane bagasse and pine wood chips [7], among others. When small amounts of agroindustrial residues are added to contaminated soil they confer soil apparent bulk density and porosity, help to diffuse oxygen between the particles and increase water retention. They are also used to support the growth of exogenous microorganisms, which are bioaugmented in soil to accelerate the degradation process, and, because of their nature, serve as carbon, phosphorus and nitrogen sources which are potentially important for the growth of organic pollutant degrading microorganisms [8]. Agroindustrial waste also contributes microorganisms with the ability to degrade toxic compounds; some studies, for example, have demonstrated that microbial biostimulation in a soil/sugarcane bagasse system at a ratio of 85:15 could remove 74% of total petroleum hydrocarbons (TPH) from the soil at 16 days [9, 10].

Several ligninolytic fungi have been grown on sugarcane bagasse and used as inoculum for the bioremediation of soil contaminated with polychlorinated biphenyls [10], phenanthrene [11], and benzo(a)pyrene [12]; such lignocellulosic materials are the natural habitat of the fungi. Previous work [13] has reported that non-ligninolytic filamentous fungi, such as *Aspergillus niger* and *Penicillium frequentans*, grown on sugarcane bagasse and added to a soil spiked with 400 ppm phenanthrene, achieved 54% removal of the pollutant from the soil after 7 days, while a mixed culture of *P. frequentans* and *Pseudomonas pickettii* achieved 73.6% removal at 18 days [11].

Filamentous fungi offer certain advantages over bacteria for bioremediation in solid culture because of their rapid colonization of solid substrates, such as soil or agroindustrial residues. In addition, they secrete large numbers of extracellular enzymes in solid culture and tolerate high concentrations of toxic compounds [14].

The most extensive studies have focused on white-rot basidiomycetes species such as *P. chrysosporium*, *Pleurotus ostreatus*, and *Trametes versicolor*. These microorganisms degrade PAHs cometabolically. The removal of PAHs by ligninolytic fungi has been attributed mainly to their extracellular ligninolytic enzymes [15-18], but their preference to colonize compact wood is a clear disadvantage since it limits their capability to grow in a completely different environment such as soil [19-22].

Also, non-ligninolytic fungi, such as *Cunninghamella elegans*, *Penicillium janthinellum*, *Aspergillus niger* and *Syncephalastrum sp.*, are able to transform a variety of PAHs, including chrysene and benzo(a)pyrene, to polar metabolites [13, 14, 23, 24]. These microorganisms carry out a mono-oxygenation of the PAH molecules by the intracellular cytochrome P-450 dependent oxidase system [25]. These fungi do not produce extracellular peroxidases, however, they do produce cytochrome P450 monooxygenase which can oxidize PAHs to epoxides and dihydrodiols: highly potent carcinogens that accumulate in soil (figure 1) [26-29].

The efficient application in bioremediation of contaminated soils is dependent, then, on having fungal strains with the ability to grow in contaminated soil without being displaced by indigenous microflora and which also produce efficient PAH-degrading enzymes such as lignin and manganese peroxidases or phenoloxidases which allow the mineralization of toxic compounds (figure 1).

To achieve this goal, genetic engineering has been an important tool to generate genetically modified microorganisms (GEMs) through the expression of gene clusters encoding the degradation of a wide variety of pollutants. For example, simple aromatics, nitro aromatics, chloroaromatics, polycyclic aromatics, biphenyls, polychlorinated biphenyls, oil components etc., have been cloned and characterized for an increased degradation potential compared to their naturally occurring counterparts. Studies have focused primarily on bacteria and obtained good results for bioremediation systems [30, 31]. Knowledge of similar activities in fungi is limited to some white-rot fungi and a few species of non-ligninolytic fungi; however some studies have focused on toxic compound degradation, where recombinant strains were more efficient in the removal of PAHs from soil than wild-type strain [32]. It is therefore hypothesized that heterologous expression of genes codifying MnP and LiP in non-ligninolytic fungi will complement the degradation pathway of cytochrome P450 to obtain complete mineralization of the hydrocarbon without leaving more toxic intermediary compounds which accumulate in the soil (figure 1).

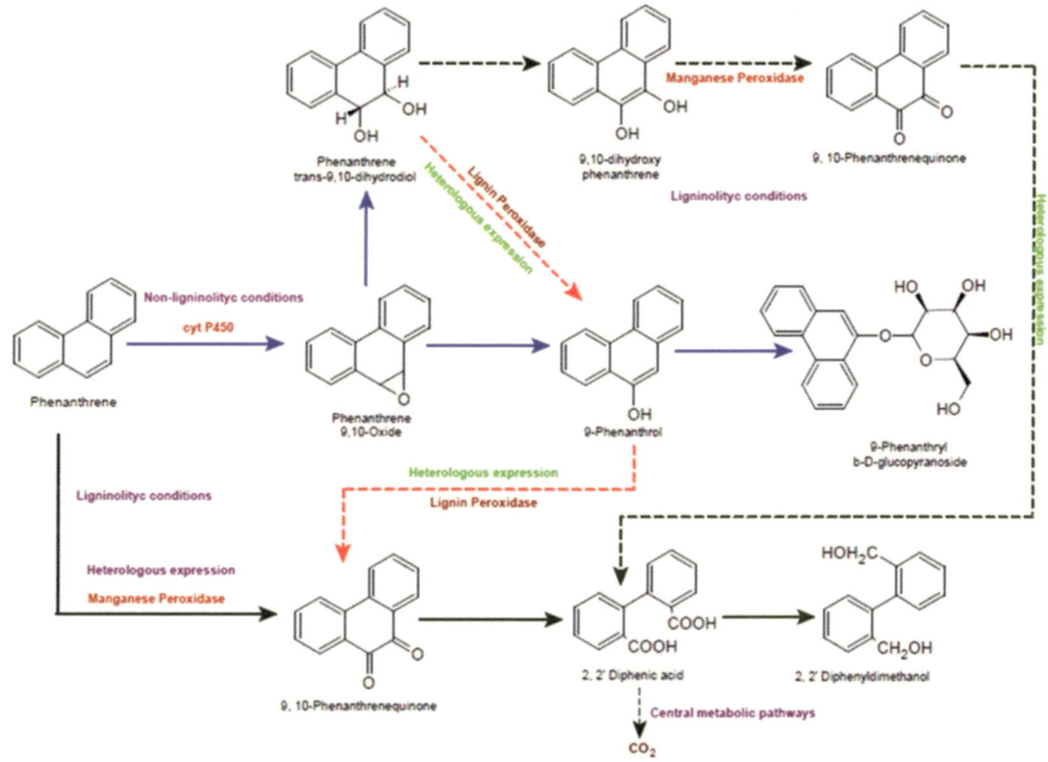

Figure 1. A proposed or hypothetical metabolic pathway for Phe degradation under ligninolytic and non-ligninolytic conditions and possible changes in the degradation pathway for the heterologous expression of genes encoding the production of peroxidase enzymes in non-ligninolytic fungi.

Some studies on the homologous expression of peroxidases in ligninolytic fungi in submerged culture, have shown that a transformed *Pleurotus ostreatus* was better at removing recalcitrant pollutant than the wild-type strain [33]. Other researchers reported the development of a homologous expression system for MnP and LiP in *P. chrysosporium*. The constitutively expressed *P. chrysosporium* glyceraldehyde phosphate dehydrogenase (*gpd*) promoter was used to drive the expression of the recombinant genes, now using nutrient-rich media in which the endogenous genes are not expressed. However, and despite the use of the strong promoter, production levels of the recombinant proteins remained at the same low level as is normally produced by the endogenous genes under starvation conditions [34, 35].

A study on the heterologous expression of these genes has been done on the baculovirus expression system [36]. In the *E. coli* expression system, LiPH8 was found to be expressed in inactive inclusion bodies although activation was done in vitro [37], and the heterologous expression of MnP H4 isoenzyme in *E. coli* was also demonstrated [38]. The capacity of non-ligninolytic fungi to remove toxic compounds in solid culture may be used for the heterologous expression of peroxidase-encoding genes (manganese peroxidase, MnP and lignin peroxidase, LiP) from *P. chrysosporium*, thus increasing the degradation potential of PAHs in solid culture for soil bioremediation.

The expression of the lignin peroxidase gene of *Phlebia radiata* in *Trichoderma reesei*, failed to produce any extracellular LiP [39]. Overexpression of the *P. chrysosporium mnp*1 gene has been previously achieved in *Aspergillus oryzae* but at levels similar to that of the parental host in liquid culture [40]. Gene expression of *lip*A in *A. niger* F38 has also been studied, however LiP H8 activity was still detected at low levels [41]. Other researchers have used the protease-deficient *A. niger* strain for the expression of *mnp*1 and *lip*A [42]. Several factors have been identified which hamper the overproduction of recombinant proteins in filamentous fungi. In the case of heme-containing proteins, for example, limited heme availability has been indicated as a limiting factor [43, 44]. None of these studies focused on recombinant enzyme production in solid culture or its application in bioremediation systems.

We have studied the possibility of producing these peroxidases in non-ligninolytic fungi isolated from contaminated soil because of their capacity to remove PAHs in soil; a number of filamentous fungal species are capable of secreting large amounts of proteins into the medium.

In our laboratory, one fungal strain was isolated from sugarcane bagasse using Mexican "Mayan" crude oil as carbon source [9]. This strain was identified by the sequence of ITS (Internal Transcription Spacer) fragments as: *Aspergillus niger* SCB2. The strain was able to tolerate (800 ppm) and to remove 45% of the initial Phe in solid culture, using sugarcane bagasse as texturizer with Phe-contaminated soil [13].

Aspergillus niger SCB2 was used to express a manganese peroxidase gene (*mnp*1) from *P. chrysosporium* using the inducible Taka amylase promoter and secretion signal from *A. oryzae* and the glucoamylase terminator of *A. awamori* [40], aiming at increasing its PAH degradation capacity. Transformants were selected based on their resistance to hygromycin B and the discoloration induced on Poly R-478 dye by peroxidase activity. The kinetics of *A. niger* SCB2-T3 were measured in complete medium supplemented with hemoglobin to increase the MnP activity. No MnP activity was detected for the wild-type strain; however, the transformant strain of *A. niger* showed higher enzymatic activity in the presence of hemoglobin. The maximum specific activity of the SCB2-T3 strain was 3 U/l, whereas the control strains of *P. chrysosporium* reached 7.8 U/l. The maximum activity was obtained at 72 h for transformant and control strains. The transformants presented activity starting at 24 h, whereas the control strain presented maximum activity only at 72 h. In solid culture the recombinant *A. niger* SBC2-T3 strain was able to remove 95% of the initial Phe (400 ppm) from a microcosm soil system after 17 days, whereas the wild strain removed 72% under the same conditions. [32].

Although the transformant SCB2-T3 strain presented MnP enzymatic activity and production was maintained for 5 d, production levels of the recombinant proteins still remained lower than the control strain of *P. chrysosporium*, possibly because the promoter used for *mnp*1 expression was a maltose-inducible promoter. We, therefore, chose to analyze *mnp*1 expression regulated by a constitutive strong promoter (glyceraldehyde phosphate dehydrogenase, *gpd*A of *A. nidulans*) in *A. niger* SCB2 strain, and show the effect of heterologous expression of *mnp*1 gene in the transformant strain on the removal of Phe in solid culture using sugarcane bagasse as texturizer at microcosm level, and compare its degradation effectiveness with the wild-type strain.

In this study we obtained an effective heterologous expression of the mnp1 cassette controlled by the gpdA constitutive promoter in A. niger SCB2 strain. The MnP+7 transformant strain was selected due to its mayor MnP enzymatic activity after 48 h culture and up to 7 d, this important result shows that the new promoter favors protein production with catalytic activity from growth to idiophase, in comparison with SCB2-T3 strain, which shows only recombinant enzyme production while there was maltose in the culture medium, the compound that induces mnp1 gene expression in this strain. It is important for bioremediation systems that the oxidation involved PAHs enzymes are produced while stay in the soil. The longer the time in soil the increased enzyme production and higher removal of toxic compounds. On the other hand, the MnP+7 strain, was able to grow, tolerates and efficiently removed high Phe concentrations in contaminated soil as compared with the wild-type strain. After heterologous expression and the acquisition of these characteristics the MnP+7 strain, is a viable and important alternative for application in bioremediation of PAHs contaminated soils. This strain may have some potential as a bioaugmentation agent: is an efficient degrader of PAHs in high concentrations compared to other non-ligninolytic fungal strains which produce more toxic intermediaries than the original compound and tolerate lower PAH concentrations, and also compared to ligninolytic fungi not grown in soil and which are displaced by native soil microflora.

2. Methods

2.1. Heterologous expression of a *P. chrysosporium mnp*1 gene in *A. niger* SCB2

Aspergillus niger SCB2 [13] was used as the recipient in transformation experiments. pGMG-Hyg (Figure 2) was constructed by fusing the mature MnP cDNA to a 2326-bp fragment of the endogenous *gpd*A promoter and secretion signal from *A. oryzae* and to a 199-bp fragment containing the glucoamylase terminator of *A. awamori*. Fusions were created using T4 ligase (Fermentas). Plasmid pTAAMnP1 contains the secretion signal, MnP cDNA of *P. chrysosporium* and terminator, and was obtained from Dr. Daniel Cullen from the University of Wisconsin, USA [40]. Plasmid pDLAM89d contains the HygB resistance gene for fungi; this plasmid was donated by Dr. Jesus Aguirre of the Instituto de Fisiología Celular, UNAM, México. Plasmid pAN52.1 contains the *gpd*A promoter of *A. nidulans*.

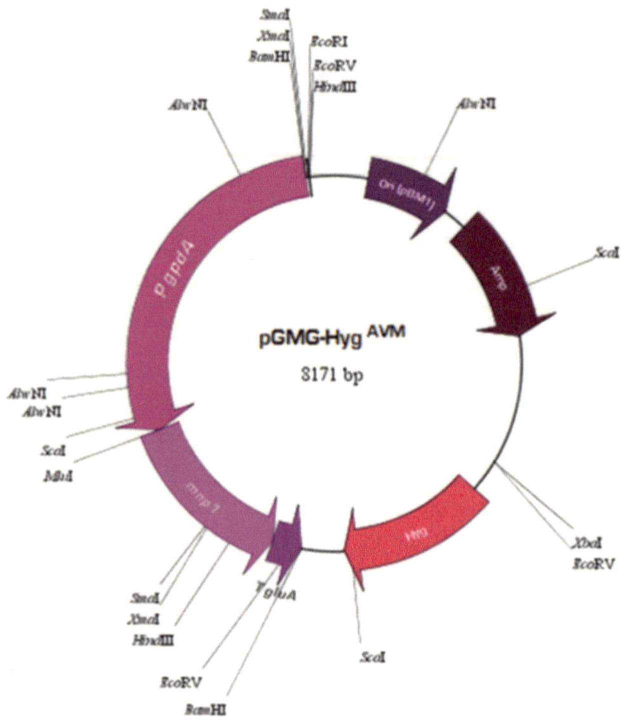

Figure 2. Expression vector pGMG-Hyg. The expression cassette contains *mnp*1 cDNA with constitutive *gpd*A promoter with the signal sequence for protein secretion and the glucoamylase gene from *A. awamori* as terminator. The plasmid contains the HygB resistance gene for the selection of fungal transformants.

Fungal transformation was done through a biolistic transformation protocol previously described for *A. nidulans* [45]. Intact conidia were inoculated into 15 ml of solid Czapek medium and then incubated for 6 h. Gold particles (1 mm diameter) were prepared and coated with plasmid pGMG-Hyg [46] and 8 μl loaded onto each of three macrocarrier discs for bombardment at 6 cm target distance, 28 in Hg vacuum and bombardment pressure of 1,200 psi. The plates were bombarded and incubated at 30°C for 2 hours, after which time a top dressing was applied of 10 ml Czapek-agar medium with 300 μg/ml hygromycin. Plates were incubated at 30°C until growth was observed. Control plates were bombarded with gold particles which were prepared as described above but not coated with plasmid.

2.2. Evaluation of enzymatic activity of recombinant MnP

Colony transformants were assayed for MnP activity using a modified plate assay method [34]. The spores of transformants obtained with HygB were inoculated onto disks (0.5 mm in diameter) of MM agar medium [47] supplemented with hemoglobin (1 g/l). The disks were incubated at 30°C for 2 d; when fungal growth began, the disks were inoculated in Petri dishes with MM agar medium, in addition to o-anisidine. The plates were incubated at 30°C for 24 h and then flooded with a solution of 50 mM Na-phosphate buffer (pH 4.5) and 0.04%

H_2O_2 on the surface of the plate and incubated at 30°C for 10 d in the dark. Positive controls were prepared with *P. chrysosporium* mycelia because this strain produces MnP; plates with the parental strain of *A. niger* were prepared for the negative control. Transformants which developed a purple halo were selected.

Incorporation of the recombinant *mnp*1 was checked through specific amplification in a PCR experiment using primers MnP1R_5173 (5′-GGATCCCTGTCTGGTCTTCTACAC-3′) and SS-MnP-MluI (5'- CGCGTATGATGGTCGCGTGGTGGTCTCT-3'). Genomic DNA was extracted from lyophilized mycelia of the wild-type and selected transformant strains, according to the described modified method [48]. After amplification, each PCR product was analyzed by agarose gel electrophoresis.

The MnP extracellular activity was determined spectrophotometrically by a modification of the method previously described using phenol red oxidation [49]. Absorbance was read at 610 nm. One unit of MnP activity was defined as 1 μmol product formed per minute.

The kinetics of wild-type and transformant strains of *A. niger* in liquid cultures were tested. Erlenmeyer flasks containing 50 ml of MM medium with hemoglobin were inoculated with 1×10^7 spores/ml and incubated at 30°C and 200 rpm for 7 d. Every 24 h, flasks were used to determine MnP activity in the supernatant. In addition, total protein concentration was quantified by Bradford reagent to determinate specific activity.

2.3. Phenanthrene removal by *A. niger* Mnp^{+7} and wild-type strain in solid culture

The ability of wild-type and transformant strains to remove Phe was determined at several times in the solid-state microcosm system, using the same culture conditions. Sugarcane bagasse was used as a fungal growth support and carbon source. The sterile material was moistened with MM medium and inoculated with 2×10^8 fungal spores/ml; all cultures were incubated for 2 d at 30°C. Uncontaminated soil obtained from a zone near a contaminated region in Coatzacoalcos, Veracruz, Mexico, was sterilized and contaminated with 600 ppm of Phe. The newly contaminated soil was mixed with the inoculated sugarcane bagasse and incubated at 30°C for 14 d, as well as a control (sterile bagasse and contaminated soil without fungi) to determine abiotic Phe removal. Evolution of CO_2 was measured daily to quantify the heterotrophic activity. After this period, Phe removal for both strains was determined by HPLC.

Heterotrophic activity was determined by Gas Chromatography (GC). Headspace samples were taken from the flasks and analyzed for CO_2 evolution. The headspace in each flask was flushed out daily for 15 min with sterile and moistened air. This allowed the preservation of aerobic conditions and avoided carbon dioxide accumulation. CO_2 quantification was reported as milligrams of CO_2 per gram of initial dry matter (IDM). Phenanthrene was extracted with microwave assisted extraction, according to EPA method 3546. Analysis of Phe was based on EPA method 3540 for the HPLC system.

3. Results

3.1. Heterologous expression of a *P. chrysosporium mnp*1 gene in *A. niger* SCB2

The heterologous expression of genes coding for different isoforms of MnP from *P. chrysosporium* has been reported in non-ligninolytic fungi, such as *A. oryzae, A. nidulans,* and *A. niger*. These enzymes were obtained extracellularly and with catalytic activity but have only been studied as expression systems for the production of heterologous proteins and have not been applied to the bioremediation of contaminated soils, as presented in this research [40,42,50,51].

Expression plasmid pGMG-Hyg was introduced into wild-type *A. niger* by biolistic transformation. A total of 8 transformants were isolated for their capacity to grow in Czapek plates with hygromycin B (HygB). After 5 d of incubation at 30°C and confirmed by PCR amplification, the result showed a single amplicon of 635 pb fragment observed in agarose gel electrophoresis (figure. 3) and no bands were observed for the wild-type stain. The positive control strain showed the expected fragment.

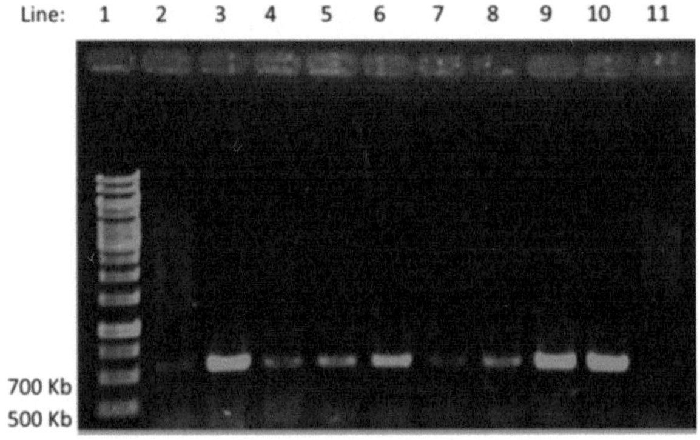

Figure 3. Agarose gel with PCR product obtained by *A. niger* transformants for expression plasmid pGMG-Hyg. Line 1: Gene Ruler 1 kb DNA ladder as a molecular marker. Line 2: MnP^{+1}. Line 3: MnP^{+2}. Line 4: MnP^{+3}. Line 5: MnP^{+4}. Line 6: MnP^{+5}. Line 7: MnP^{+6}. Line 8: MnP^{+7}. Line 9: MnP^{+8}. Line 10: *P. chrysosporium* as positive control. Line 11: *A. niger* SCB2 as negative control.

3.2. Evaluation of enzymatic activity of recombinant MnP

The transformants were evaluated for MnP activity by the o-anisidine coloration method in MM medium with hemoglobin plates. Transformants that developed a purple halo were selected. Four transformants (MnP^{+1}, MnP^{+4}, MnP^{+7} and MnP^{+8}) formed purple halos around the agar disk after 8 d of incubation, indicating extracellular peroxidase activity, as show in figure 4. The wild-type strain showed no coloration; however, the control strain of *P. chrysosporium* showed a greater purple halo. These results are similar to findings by other authors, who screened autochthonous or recombinant fungal strains for the formation of halos by o-anisidine oxidation on agar plates induced by manganese peroxidase activity [40, 42, 52].

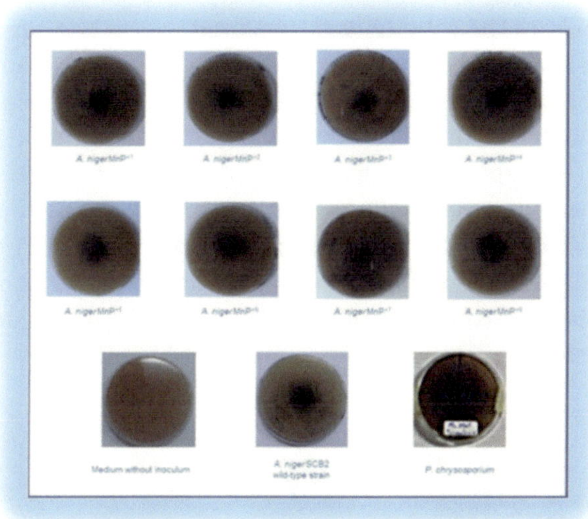

Figure 4. Qualitative determination of MnP activity produced by *A. niger* transformants in Petri dishes using o-anisidine as an indicator.

The MnP^{+7} transformant strain showed higher Phe tolerance than wild-type strain when inoculated into Cove's medium in Petri dishes at different Phe concentrations. At concentrations above 600 ppm, both strains showed a decrease in growth rate compared to their respective controls without Phe; however, the wild-type strain showed an inhibition in sporulation while the transformant strain was able to sporulate (figure 5). This coincides with the results in reference [53], which reported that fluorene at concentrations above 100 ppm caused growth inhibition of fungal strains isolated from a contaminated soil. In contrast, reference [54] reported that 100 ppm of anthracene had no inhibitory effect on the growth of fungi isolated from soil. This fact suggests that due to the production of MnP by the transformant strain for *mnp*1 gene expression, which is regulated by the *gpd*A constitutive promoter, there was an increased tolerance to high Phe concentrations.

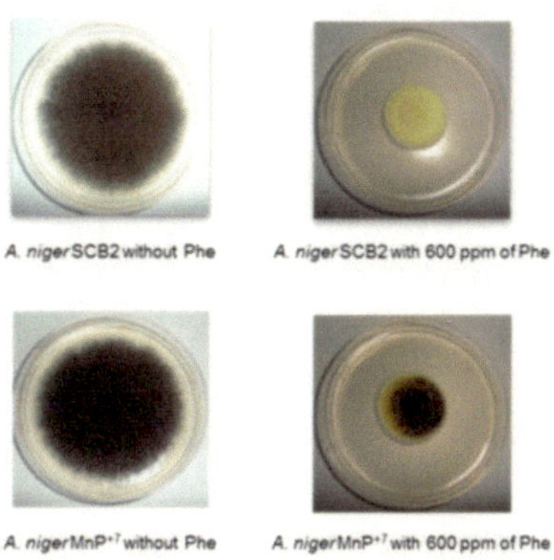

Figure 5. Effect of Phe on growth and sporulation of *A. niger* wild-type strain and transformant on superficial culture in Petri dishes with Cove's medium.

MnP productivity of the four selected strains was quantified in liquid culture using MM medium with hemoglobin. The activity was measured every 24 h for 7 days. As shown in figure 6, the wild-type strain did not present MnP activity. Although all transformant strains present different MnP activity, maximum activity was obtained by *A. niger* MnP^{+7} strain at 4 d (25.4 U/L); specific activity during this time was 3.67 U/mg of total protein, whereas the control strains of *P. chrysosporium* reached 7 U/L. These results demonstrate that the o-anisidine plate assay was consistent with the MnP production of the transformants in liquid culture. The results also confirmed that MnP production occurred in the transformants by introducing the *mnp*1 gene, since this activity was not detected in the wild-type strain. Differences observed between transformants are often explained by a differential integration of the heterologous gene within the fungal genome [33, 55].

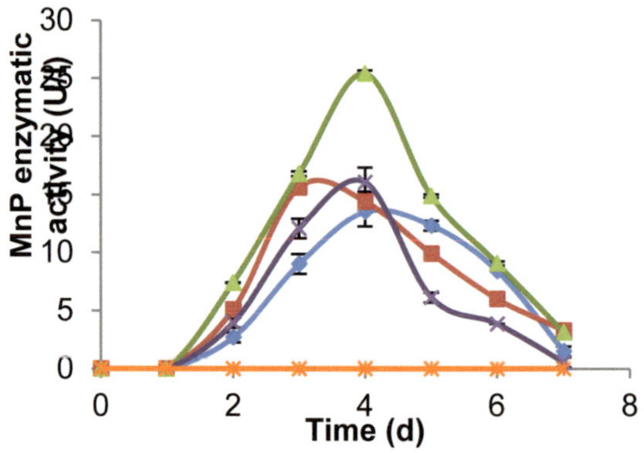

Figure 6. Enzymatic activity of MnP exerted by the wild-type strain and transformants of *A. niger* in liquid culture in MM medium supplemented with hemoglobin. (◆) *A. niger* MnP^{+1}; (■) *A. niger* MnP^{+4}; (▲) *A. niger* MnP^{+7}; (×) *A. niger* MnP^{+8}; and (✱) *A. niger* SCB2.

After 48 h, all transformant strains showed MnP activity in liquid medium. Because *mnp*1 expression is regulated by the constitutive *gpd*A promoter, which is involved in the glycolysis metabolic pathway forming part of the central metabolism, MnP production was started early on and maintained throughout the growth phase of the fungus [56].

The specific activity obtained by recombinant MnP with *A. niger* MnP^{+7} strain regulated by the constitutive *gpd*A promoter, and *A. niger* SCB2 T3 strain regulated by Taka-amylase promoter, showed similar results (3.67 and 3 U/L respectively); this may be caused by the culture media used for producing the recombinant enzymes since the medium used for SCB2 -T3 was COVE's medium, with maltose (50 g/L) as carbon source and inducer of *mnp*1 expression, supplemented with hemoglobin to increase MnP activity [32], and the medium used for MnP^{+7} was MM [47] also supplemented with hemoglobin but using glucose (10 g/L) as carbon source. It is well known that several factors should be considered in *mnp*1 expression, such as the carbon source, the culture medium, and the addition of a heme group. Recombinant MnP production by MnP^{+7} strain was increased by adding hemoglobin to the culture medium. Similar results have been obtained by other research groups in their studies on *mnp*1 expression in *Aspergillus* [40,42,51] and are explained by how the recombinant protein is produced by *Aspergillus*, i.e., as an unstable apoprotein. This apoprotein requires the heme group to produce the active hemoprotein, which acquires a more stable conformation for proteolytic attack than the apoprotein. Due to low availability of heme provided by the heme biosynthetic pathway, this fact is considered a limiting factor for the production of heme proteins in different expression systems [41- 43, 52]. If the amount of heme produced by a microorganism is low in relation to the amount of apoprotein produced, the apoprotein will accumulate in the culture medium and undergo proteolytic degradation lowering the yield of the active hemoprotein. Other studies on manganese peroxidase expression in *Aspergillus* assume that hemoglobin may play a role not only in supplying heme but also in providing a protein excess in the culture medium, thereby protecting recombinant MnP from proteolytic degradation [42,51,57].

3.3. Phenanthrene removal by *A. niger* Mnp^{+7} and wild-type strain in solid culture

In order to evaluate the growth of the microorganism in solid culture, CO_2 evolution was quantified. Two tested strains showed different profiles and the ANOVA test indicated a significant ($p<0.05$) difference in the accumulated CO_2 production: the transformant strain produced more CO_2 than the wild-type, both in the presence and absence of Phe, and was around 15-18 mg CO_2 accumulated per gram of initial dry matter (IDM), whereas the wild-type strain produced only around 5-7 mg CO_2 accumulated/g IDM (figure 7A). This result demonstrated that *A. niger* MnP^{+7} strain was able to grow more on solid culture than *A. niger* SCB2; however, both strains presented a decrease in CO_2 production in the presence of Phe. This can be interpreted as a toxic effect of the compound on the growth of fungi and these results demonstrate that the plate assay with 600 ppm of Phe was consistent with a toxic effect on the growth. The highest microbial activity in all treatments analyzed was at 4 d (5 mg CO_2 instantaneous/g IDM) and a decrease in microbial activity was observed in all the inoculum treatments at 14 days (figure 7B). CO_2 production in the control without fungi was negligible.

The residual Phe extracted from treated soil was quantified by HPLC and the results of two strains are presented in figure 8. The wild-type strain had the lowest Phe removal capacity (approximately 7%) compared with the transformant MnP^{+7} strain which was able to remove approximately 44% of the initial Phe (0.6 mg/g IMD) in 14 d. The Phe extraction efficiency of the abiotic controls was 98%.

The increase in the removal percentage of Phe by the MnP^{+7} transformant strain in solid culture suggests that it is due to the production of MnP enzyme by the transformant strain which showed the ability to express the *mnp*1 gene regulated by the constitutive *gpd*A promoter. This fact has led to an increased tolerance in plate and solid culture and greater removal efficiency in high Phe concentrations in solid culture. This result is important because PAH degradation has only been studied in submerged culture by ligninolytic fungi isolated from contaminated soils and, since PAHs have low solubility in water, only low concentrations have been used; for example, a strain of *Aspergillus terreus* has been isolated from a PAH-polluted soil and the metabolism of pyrene and benzo(a)pyrene by this fungus was investigated in liquid submerged culture supplemented with 50 and 25 ppm, respectively, of each compound [58]. *Penicillium chrysogenum* degraded 60% of fluorene (50 ppm initial) in the presence of Tween 80 after 2 days [53]. There are also reports of PAH removal by ligninolytic fungi in soils, but only concentrations below 200 ppm have been tested. *Fusarium sp.* E033 strain was isolated and able to survive in the presence of concentrations up to 300 ppm of benzo(a)pyrene and demonstrated that this strain was able to degrade 65 to 70% of the initial benzo(a)pyrene (using 100 ppm) provided within 30 d of incubation at 32°C [59]. Other authors report PAH degradation by some strains of the genus *Penicillium*, such as *P. frequentans*, capable of removing 52% of Phe in soil contaminated with 200 ppm in 17 d [60]. *P. janthinellum* degraded 50 ppm of benzo(a)pyrene after 48 d of incubation in soil in co-culture with bacteria [24].

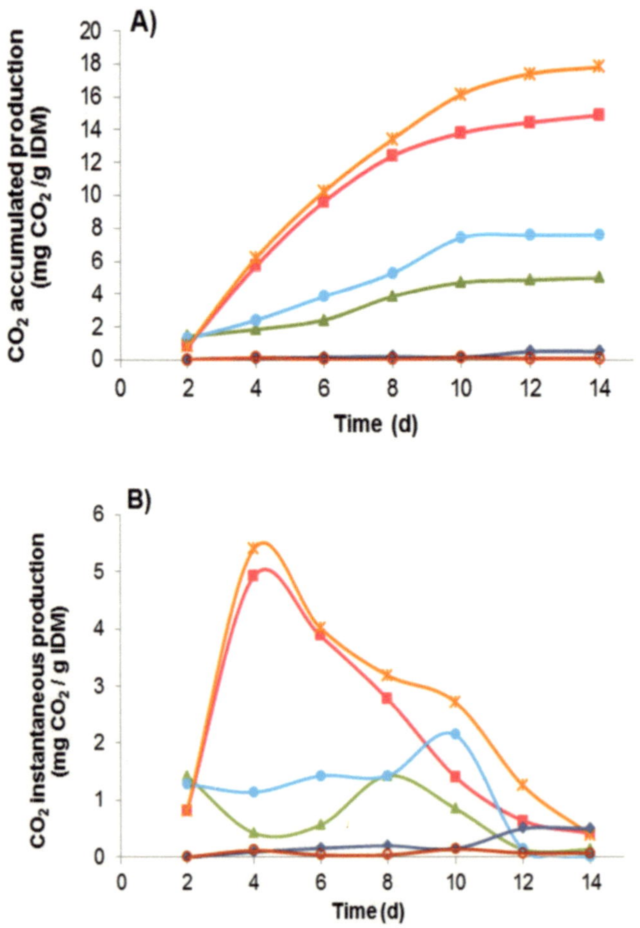

Figure 7. Microbial activity of both *A. niger* strains in solid culture in microcosm. A) CO_2 accumulated production and B) CO_2 instantaneous production. The different treatments were: (●) *A. niger* SCB2 without Phe; (▲) *A. niger* SCB2 with Phe; (✱) *A. niger* MnP^{+7} without Phe; (■) *A. niger* MnP^{+7} with Phe; (◆) abiotic control without Phe and (O) abiotic control with Phe.

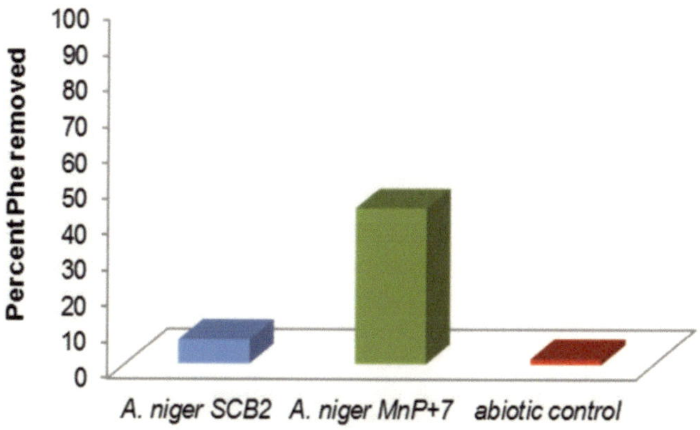

Figure 8. Phenanthrene removal percentage in solid culture by *A. niger* SCB2 and *A. niger* MnP^{+7} strains in soil contaminated with 600 ppm of Phe.

The increase in Phe concentration in solid culture showed a higher toxic effect on the wild-type strain. Compared to cultures carried out using soil contaminated with 0.4 mg/g IMD, the wild-type strain SCB2 was able to remove 75% of Phe, while *A. niger* SCB2-T3 strain removed 95% of Phe [32], so although the removal percentage obtained with MnP^{+7} strain was lower than *A. niger* SCB2-T3, MnP^{+7} strain is considered to be efficient in the removal of Phe in soil.

With respect to the intermediaries formed during Phe oxidation in solid culture for *A. niger* MnP^{+7}, preliminary results using polyurethane foam (PUF) as inert support have shown that the chromatographic profiles of the two strains in the presence of 600 ppm of Phe were different and the peaks observed in the chromatograms of the two strains in the presence of Phe were not observed in the respective controls (control inoculated without Phe and abiotic controls with Phe) (data not shown). This suggests that these peaks correspond to metabolites formed from Phe degradation; furthermore, the concentrations of residual Phe and intermediaries obtained from the transformant strain were lower when compared to those obtained from the wild-type strain. This result can be attributed to *mnp*1 expression since this is the only difference between the two strains, which were grown under the same culture conditions. These results allow us to infer that the differences in metabolism were caused by the presence of recombinant MnP enzyme, although we, have not yet identified the metabolites or intermediaries formed from degradation by transformant MnP^{+7} and wild-type strains.

1-phenanthrol, 2-phenanthrol, and phenanthrene trans-9,10-dihydrodiol have been reported as major metabolites from the metabolism of Phe by *A. niger* [61]. Other authors have reported that Phe was metabolized by *A. niger* into small amounts of 1- and 2-phenanthrol, and also 1-methoxyphenanthrene as a major ethyl acetate extractable metabolite; its retention time (RT) was of 36.7 min, indicating a less polar metabolite than Phe [62]. We have reported that the chromatographic profiles of *A. niger* SCB2-T3 in liquid culture in the presence of Phe were different to the wild-type strain. In the culture medium of the wild-type strain, a principal metabolite of Phe degradation with a RT of 1.7 min was detected. On the other hand, two metabolites with a lower RT than Phe were extracted from the mycelium of the transformant strain; these are considered more polar compounds produced by Phe oxidation. No residual Phe was detected in the transformant strain's culture medium [32].

Bioaugmentation with an *Aspergillus* strain isolate did increase the extent of removal of benzo(a)anthracene and benzo(a) pyrene in soils significantly [65]. Bioaugmentation with *Cladosporium sphaerospermum* significantly stimulated PAH degradation, especially for high molecular weight PAH [66]. Other authors have reported a significantly improved degradation of high molecular weight PAH after bioaugmentation in PAH-contaminated soil [24, 29].

The results from this study also show that non-ligninolytic fungal strains are a viable alternative for application in bioremediation systems; moreover, bioaugmentation with genetically modified exogenous fungal strains for heterologous protein production in solid culture accelerates the process of removal and biodegradation of toxic compounds in contaminated soils.

4. Conclusion

The action of genetically modified non-ligninolytic fungal strains in bioremediation systems has not been reported, so that the results obtained in this investigation suggest that these microorganisms may have some potential as a bioaugmentation agent: they are efficient degraders of PAHs in high concentrations compared to other non-ligninolytic fungal strains which produce more toxic intermediaries than the original compound and tolerate lower PAH concentrations, and also compared to ligninolytic fungi not grown in soil and which are displaced by native soil microflora.

Author details

Diana V. Cortés-Espinosa
Corresponding Author
Centro de Investigación en Biotecnología Aplicada del IPN, Tepetitla de Lardizabal, Tlaxcala, México

Ángel E. Absalón
Centro de Investigación en Biotecnología Aplicada del IPN, Tepetitla de Lardizabal, Tlaxcala, México

Acknowledgement

This work was supported by SEP-CONACYT, project CB2008-105643 and Instituto Politécnico Nacional, project SIP20121707.

5. References

[1] IARC. Monographs on the evaluation of the carcinogenic risk of chemicals to humans. Vol. 92. Some Non-heterocyclic Polycyclic Aromatic Hydrocarbons and Some Related Exposures. Lyon France. 2010.

[2] Sudip, K. S., Singh O.V. and R.K. Jain. (2002) Polycyclic aromatic hydrocarbons: environmental pollution and bioremediation, Trend. Biotechnol. 20: 243–248.

[3] Chen, B., X. Xuan, L. Zhu, J. Wang, Y. Gao, K. Yang, X. Shen and B. Lou (2004). Distributions of polycyclic aromatic hydrocarbons in surface waters, sediments and soils of Hangzhou city China, Water Res. 38: 3558–3568.

[4] Shiaris, M.P (1989) Seasonal biotransformation of naphthalene, phenanthrene and benzo(a)pyrene in surficial estuarines sediments. Appl. Environ. Microbiol. 55:1391-1399.

[5] Black, J.A., W.J. Birge, A.G. Westerman and P.C Francis (1983) Comparative aquatic toxicology of aromatic hydrocarbons. Fundamental and Appl. Toxicol, 3:353-358.

[6] White, K.L (1986) An overview of immunotoxicology and carcinogenic polycyclic aromatic hydrocarbons. Environ. Carcin. Re. C4:163-202.

[7] Sample, K.T., Reid, B.J., Fermor, T.R (2001) Review of composting strategies to treat organic pollutants in contaminated soils. Environ. Pollut. 112: 269–283.

[8] Pandey, A., Soccol, C., Nigam, P., Soccol, V (2000) Biotechnological potential of agroindustrial residues I: Sugarcane bagasse. Bioresource Technol. 74, 69–80.

[9] Pérez-Armendáriz B, Loera-Corral O, Fernández-Linares L, Esparza-García F, Rodríguez-Vázquez R (2004) Biostimulation of micro-organisms from sugarcane bagasse pith for the removal of weathered hydrocarbon from soil. Lett Appl Microbiol; 38: 373–377.

[10] Fernández-Sánchez, J.M., R. Rodríguez-Vázquez, G. Ruiz-Aguilar and P.J.J. Alvarez (2001) PCB biodegradation in aged contaminated soil: interactions between exogenous *Phanerochaete chrysosporium* and indigenous microorganisms. J. Environ. Sci. Health, part. A. 36(7):1145-1162.

[11] Chávez-Gómez, B., Quintero, R., Esparza-García, F., Mesta-Howard, A., de la Serna, F.J.Z.D., Hernández-Rodríguez, C., Gillen, T., Poggi-Varaldo, H., Barrera-Corte´ s, J., Rodríguez-Vázquez, R (2003) Removal of phenanthrene from soil by co-cultures of bacteria and fungi pregrown on sugarcane bagasse pith. Bioresource Technol. 89: 177–183.

[12] Dzul-Puc, J., Esparza-García, F., Barajas-Aceves, M., Rodríguez-Vázquez, R (2004) Benzo[a]pyrene removal from soil by *Phanerochaete chrysosporium* grown on sugarcane bagasse and pine sawdust. Chemosphere 58, 1–7.

[13] Cortés-Espinosa, D.V., Fernández-Perrino, F.J., Arana-Cuenca, A., Esparza-García, J.F., Loera, O., Rodríguez-Vázquez, R (2006) Selection and identification of fungi isolated from sugarcane bagasse and their application for phenanthrene removal from soil. J. Environ. Sci. Health., Part A. 41(3): 475–486.

[14] Cerniglia, C.E., G.L. White and R.H. Helflich (1985) Fungal metabolism and detoxification of polycyclic aromatic hydrocarbons. Arch. Microbiol. 143:105-110.

[15] Barr, D.P. and S.D. Aust (1994) Mechanisms white rot fungi use to degrade pollutants. Environ Sci. technol. 28:79-87.

[16] Bezalel, L, Y. Hadar, P.P. Fu, J.P. Freeman and C.E. Cerniglia (1996) Initial oxidation products in the metabolism of pyrene, anthracene, fluorene and dibenzothiophene by the white rot fungus *Pleurotus ostreatus*. Appl Environ Microbiol. 62:2554-2559.

[17] Bezalel, L, Y. Hadar, P.P. Fu, J.P. Freeman and C.E. Cerniglia. 1996. Metabolism of phenantrene by the white rot fungus *Pleurotus ostreatus*. Appl. Environ. Microbiol. 62:2547-2553.

[18] D´Annibale, A., M. Ricc, V. Leonardi, D. Quaratino, E. Mincione and M. Petruccioli (2005) Degradation of aromatic hydrocarbons by white-rot fungi in a historically contaminated soil. Biotechnol. Bioeng. 90:723-731.
[19] Boyle, C.D (1995) Development of a practical method for inducing white rot fungi to grow into and degrade organopollutants in soil. Can. J. Microbiol. 41: 345–353.
[20] Capotorti, G., P. Digianvincenzo, P. Cesti, A. Bernardi and G. Guglielmetti (2004) Pyrene and benzo(a)pyrene metabolism by an *Aspergillus terreus* strain isolated from a polycyclic aromatic hydrocarbons polluted soil. Biodegradation 15: 79–85.
[21] Meysami, P. and H. Baheri (2003) Pre-screening of fungi and bulking agents for contaminated soil bioremediation. Adv Environ Res. 7:881–887.
[22] Okeke, R.C. and H.U. Agbo (1996) Influence of environmental parameters on pentachlorophenol biotransformation in soil by *Lentinus edodes* and *Phanerochaete chrysosporium*. Appl. Microbiol. Biotechnol. 45:263–266.
[23] Launen, L., Pinto, L., Wiebe, C., Kielmann, E. and Moore, M (1995) The oxidation of pyrene and benzo(a)pyrene by nonbasidiomycete soil fungi. Can. J. Microbiol. 41:477–488.
[24] Boonchan S, Britz ML, Stanley GA (2000) Degradation and mineralization of high-molecular-weight polycyclic aromatic hydrocarbons by defined fungal-bacterial cocultures. Appl Environ Microbiol. 66:1007–1019.
[25] Gramss G, Voigt KD & Kirsche B (1999) Degradation of polycyclic aromatic hydrocarbons with three to seven aromatic rings by higher fungi in sterile and unsterile soils. Biodegradation 10:51–62
[26] Shimada, T. (2006). Xenobiotic-metabolizing enzymes involved in activation and detoxification of carcinogenic polycyclic aromatic hydrocarbons. Drug. Met. Pharmacokinet. 21(4):257-276.
[27] Shimada, T. and Y. Fujii-Kuriyama (2004) Metabolic activation of polycyclic aromatic hydrocarbons to carcinogens by cytochromes P450 1A1 and 1B1. Cancer Sci. 95(1):1–6
[28] Sutherland, J.B (1992) Detoxification of polycyclic aromatic hydrocarbons by fungi. J. of Ind. Microbiol. 9(1): 53-62.
[29] Juhasz, A. L. and R. Naidu (2000) Bioremediation of high molecular weight polycyclic aromatic hydrocarbons: A Review of the Microbial Degradation of Benzo[a]pyrene. Int. Biodeterior. Biodegrad. 45: 57-88.
[30] Pieper, D.H. and W. Reineke (2000) Engineering bacteria for bioremediation. Curr. Opin. Biotechnol. 11:262–270.
[31] Sayler, G.S. and S. Ripp (2000) Field applications of genetically engineered microorganisms for bioremediation processes. Curr. Opin. Biotechnol. 11:286–289.
[32] Diana V. Cortés-Espinosa, Ángel E. Absalón, Noé Sánchez, Octavio Loera, Refugio Rodríguez-Vázquez and Francisco J. Fernández (2011) Heterologous expression of manganese peroxidase in *Aspergillus niger* and its effect on phenanthrene removal from soil. J Mol Microbiol Biotechnol. 21:120–129
[33] Tsukihara T., Y. Honda, T. Watanabe and T. Watanabe (2006) Molecular breeding of white rot fungus *Pleurotus ostreatus* by homologous expression of its versatile peroxidase MnP2. Appl Microbiol Biotechnol. 71:114–120.
[34] Mayfield, M. B., K. Kishi, M. Alic, and M. H. Gold (1994) Homologous expression of recombinant manganese peroxidase in *Phanerochaete chrysosporium*. Appl. Environ. Microbiol. 60:4303–4309.

[35] Sollewijn Gelpke, M. D., M. Mayfield-Gambill, G. P. Lin Cereghino, and M. H. Gold (1999) Homologous expression of recombinant lignin peroxidase in *Phanerochaete chrysosporium*. Appl. Environ. Microbiol. 65:1670–1674.

[36] Johnson T.M. and J.K. Li (1991) Heterologous expression and characterization of an active lignin peroxidase from *Phanerochaete chrysosporium* using recombinant baculovirus. Arch Biochem Biophys. 291(2):371:378.

[37] Doyle W. and A.T. Smith (1996) Expression of lignin peroxidase H8 in *Escherichia coli*: folding and activation of the recombinant enzyme with Ca^{2+} and haem. Biochem J. 315:15-19.

[38] Whitwam R. and M. Tien (1996) Heterologous of fungal Mn peroxidase in *E. coli* and refolding to yield active enzyme. Biochem Biophys Res Commun 216:1013-1017.

[39] Saloheimo, M., V. Barajas, M.-L. Niku-Paavola, and J. K. C. Knowles (1989) A lignin peroxidase-encoding cDNA from the white fungus *Phlebia radiata*: characterization and expression in *Trichoderma reesei*. Gene 85:343–351.

[40] Stewart, P., R. Whitwam, P. Kersten, D. Cullen and M. Tien (1996) Efficient expression of *Phanerochaete chrysosporium* manganese peroxidase in *Aspergillus oryzae*. Appl. Environ. Microbiol. 62: 860-864.

[41] Aifa, M.S, S. Sayadi and A. Gargouri (1999) Heterologous expression of lignin peroxidase of *Phanerochaete chrysosporium* in *Aspergillus niger*. Biotechnol. lett. 21:849-853.

[42] Conesa, A., C. A. Van Den Hondel, and P. J. Put (2000) Studies on the production of fungal peroxidases in *Aspegillus niger*. Appl. Environ. Microbiol. 66(7): 3016-3023.

[43] Andersen, H. D., E. B. Jensen, and K. G. Welinder. September (1992) A process for producing heme proteins. European Patent Application EP 0 505 311 A2.

[44] Fowler, T., M. W. Rey, P. Vaha-Vahe, S. D. Power, and R. M. Berka (1993) The catR gene encoding a catalase from *Aspergillus niger*: primary structure and elevated expression through increased gene copy number and use of a strong promoter. Mol. Microbiol. 9:989–998.

[45] Fungaro, M. H. P., Rech, E., Muhlen, G. S., Vainstein, M. H., Pascon, R. C., deQueiroz, M. V., Pizzirani-Kleiner, A. A. and de Azevedo J. L (1995) Transformation of *Aspergillus nidulans* by microprojectile bombardment on intact conidia, FEMS Microbiol. Lett. 125: 293-298

[46] Herzog, R.W., Daniell, H., Singh, N.K. y Lemke, P.A. (1996) A comparative study on the transformation of *Aspergillus nidulans* by microprojectile bombardment of conidia and a more conventional procedure using protoplasts treated with polyethyleneglycol. Appl. Microbiol. Biotechnol. 45:333- 337

[47] Bennett, J.W., Lasure, L.L (1991) Growth media. In: Bennett, J.W., Lasure, L.L. (Eds.), More Gene Manipulations in Fungi. Academic Press, San Diego, pp. 441-458.

[48] Reader U, Broda P (1985) Rapid preparation of DNA from filamentous fungi. Lett. Appl. Microbiol. 1:17–20.

[49] Kuwahara M, Glenn JK, Morgan MA, Gold MH (1984) Separation and characterization of two extracellular H_2O_2 dependent oxidases from ligninolytic culture of *Phanerochaete chysosporium*. FEBS Lett. 169: 247–250.

[50] Larrondo, L.F., S. Lobo, P. Stewart, D. Cullen and R. Vicuña (2001) Isoenzyme multiplicity and characterization of recombinant manganese peroxidase *Ceriporiopsis subvermispora* and *Phanerochaete chrysosporium*. Appl. Environ. Microbiol. 67:2070-2075.

[51] Singh, D. and S. Chen (2008) The white-rot fungus *Phanerochaete chrysosporium*: conditions for the production of lignin-degrading enzymes. Appl. Microbiol Biotechnol. 81:399-417.
[52] Lokman BC, Joosten V, Hovenkamp J, Gouka RJ, Verrips CT, van den Hondel (2003) Efficient production of *Arthromyces ramosus* peroxidase by *Aspergillus awamori*. J Biotechnol. 103: 183–190.
[53] Garon D, Krivobok S, Wouessidjewe D, Seigle-Murandi F (2002) Influence of surfactants on solubilization and fungal degradation of fluorene. Chemosphere. 47:303–309.
[54] Krivobok S., E. Miriouchkine, F. Seigle-Murandi, J.-L. Benoit-Guyod (1998) Biodegradation of Anthracene by soil fungi. Chemosphere 37(3):523-530.
[55] Nevalainen, K.M.H., V.S.J. Te o and P.L. Bergquist (2005) Heterologous protein expression in filamentous fungi. Trends Biotechnol. 23(9):469-474.
[56] Glenn, J. K., and M. H. Gold (1985) Purification and characterization of an extracellular Mn(II)-dependent peroxidase from the lignin-degrading basidiomycete *Phanerochaete chrysosporium*. Arch. Biochem. Biophys. 242:329–341.
[57] Elrod S, Zelson S (1999) Methods for increasing hemoprotein production in fungal mutants. Word Intellectual Property Organization. WO/1999/029874.
[58] Capotorti, G., P. Digianvincenzo, P. Cesti, A. Bernardi and G. Guglielmetti (2004) Pyrene and benzo(a)pyrene metabolism by an *Aspergillus terreus* strain isolated from a polycyclic aromatic hydrocarbons polluted soil. Biodegradation 15: 79–85.
[59] Chulalaksananukul S, Gadd GM, Sangvanich P, Sihanonth P, Piapukiew J, Vangnai AS (2006) Biodegradation of benzo(a)pyrene by a newly isolated *Fusarium sp*. FEMS Microbiol Lett. 262:99–106.
[60] Meléndez-Estrada J, Amezcua-Allieri MA, Alvarez PJJ, Rodríguez-Vázquez R (2006) Phenanthrene removal by *Penicillium frequentans* grown on a solid-state culture: effect of oxygen concentration. Environ Technol. 27:1073–1080.
[61] Casillas RP, Crow Jr. SA, Heinze TM, Deck J, Cerniglia CE (1996) Initial oxidative and subsequent conjugative metabolites produced during the metabolism of phenanthrene by fungi. J Ind Microbiol. 16:205–215.
[62] Sack U, Heinze TM, Deck J, Cerniglia CE, Cazau MC, Fritsche W (1997) Novel metabolites in phenantrene and pyrene transformation by *Aspergillus niger*. Appl Environ Microbiol. 63:2906–2909.
[63] Hammel KE, Gai WZ, Green B, Moen MA (1992) Oxidative degradation of phenanthrene by the ligninolytic fungus *Phanerochaete chrysosporium*. Appl Environ Microbiol. 58:1832–1838.
[64] Sutherland, JB, Selby AL, Freeman JP, Evans FE, Cerniglia CE. (1991) Metabolism of phenanthrene by *Phanerochaete chrysosporium*. Appl. Environ. Microbiol. 57:3310–3316.
[65] Serrano Silva I, da Costa dos Santos E, Ragagnin de Menezes C, Fonseca de Faria A, Franciscon E, Grossman M, Durrant LR (2009) Bioremediation of a polyaromatic hydrocarbon contaminated soil by native soil microbiota and bioaugmentation with isolated microbial consortia. Biores Technol. 100:4669–4675.
[66] Potin O, Veignie E, Rafin C (2004) Biodegradation of polycyclic aromatic hydrocarbons (PAHs) by *Cladosporium sphaerospermum* isolated from an aged PAH contaminated soil. FEMS Microbiol Ecol. 51:71–78.

Chapter 7

Characterizing Microbial Activity and Diversity of Hydrocarbon-Contaminated Sites

Aizhong Ding, Yujiao Sun, Junfeng Dou,
Lirong Cheng, Lin Jiang, Dan Zhang and Xiaohui Zhao

Additional information is available at the end of the chapter

http://dx.doi.org/10.5772/50480

1. Introduction

Hydrocarbons, one of the major petroleum constituents, mainly include saturated alkanes and cycloalkanes, unsaturated alkenes, alkynes and aromatic hydrocarbons. The usual composition of light crude oil is 78% saturates, 18% aromatics, 4% resins and <2% asphaltenes (Olah and Molnar 2003). The hydrocarbon fractions in the order of decreasing volatility are C_6- C_{10}; C_{10} – C_{16}; C_{16} - C_{34} and C_{34} - C_{50}, in which C_x is referred to the number of carbon molecules in the alkane backbone.

Polycyclic aromatic hydrocarbons (PAH) are compounds consisted of two or more fused benzene rings in linear, angular, of cluster arrangements (Johnson et al, 1985), and they are mainly associated with industrial processes, though they also occur as natural constituents of unaltered fossil fuels. By definition, only C and H atoms are shown in PAH structures, but nitrogen, sulfur and oxygen atoms may readily substitute in the benzene ring to form herterocyclic aromatic compounds, commonly grouped with PAHs. In general, petroleum, coal, by-products of industrial processing, the products of incomplete combustion of organic compounds are considered to be the major sources of PAHs. Anthropogenic sources, particularly the combusting of fossil fuels, are the most significant sources of PAHs entering to the environment (Lim, et al. 1999). Sites where refining petroleum have occurred are frequently contaminated with PAHs and other aromatic hydrocarbons (Smith, 1990). Spills and leak from the petroleum storage tanks also cause significant PAH contamination.

Exposure to PAHs constitutes a significant health risk for people living in the industrialized areas of the world. PAHs are suspected human carcinogens, and they have been linked to genotoxic, reproductive, and mutagenic effects in humans. Most PAHs have long environmental persistence, their hydrophobic nature and corresponding limited water solubility lead to be recalcitrant to biodegradation and remain as environmental contaminants for extremely long time periods. The chemical properties are related to molecule size and the pattern of ring linkage. Persistence increases as PAH molecular weight increases. Low-molecular-weight PAHs (those containing less than four benzene rings) are acutely toxic, with some having effects on the reproduction and mortality rates in aquatic animals. Most high-molecular –weight PAHs (containing four or more benzene rings) are mutagenic and carcinogenic (Boonchan et al., 2000). Therefore, the US Environmental protection Agency (USEPA) has listed 16 PAHs as priority pollutants to be monitored in industrial waste effluents.

Sites contaminated with hydrocarbons are difficult to remediate. The type of petroleum product discharged will influence the terrestrial migration; in general, the more viscous the product, the slower it will migrate. Smaller more volatile fractions will move quickly, or evaporate, while larger fractions will take more time to flow through soil. The spill profile, or spatial area directly contacted with discharged hydrocarbons, depends on the time, amount, extent, and type of petroleum product, in addition, the soil particle size of the spill site also is the important factor. Hydrocarbons move through the soil and larger soil particles generally allow greater migration. The state of the ground on which the spill is discharged affects the vertical and horizontal spill profile.

When the hydrocarbons flow down through the soil, some of the organic carbon will be removed from the system by abiotic uptake. Some of the hydrocarbons such as PAHs, can be adsorbed to humic substances. The sorption characteristics of hydrocarbons to soil can further depend on the soil matrix; in a marsh environment with multiple soil types, a greater reduction of hydrocarbons was observed in sandy soils than in mineral soils (Lin et al. 1999).

The loss of hydrocarbons by abiotic processes in a system is finite, reaching a saturation point that, unless conditions change, will prevent further removal (Ping et al. 2006). This is apparent in aged spills where the residual hydrocarbons, especially for PAHs, can remain in the soil for long time without any apparent removal by abiotic processes. Volitization of hydrocarbons can remove hydrocarbons, mostly in the C6- C10 fraction, from hydrocarbon contaminated sites, but the cold temperatures will greatly reduce this volitization.

Bioremediation that uses microorganisms to remove toxic substances from the environment in an attempt to return the environment to pre-contaminated conditions has been developing for some 40 years. Bioremediation of PAHs contaminated soil and sediment is believed to be a promising cleanup process in comparison to other remediation processes such as chemical or physical ones. Therefore, many investigations have focused attention on bioremediation process that showed higher efficiency to clean up the contaminated sites. The approaches for the bioremediation of PAHs contaminated sites have included the three major processes: monitored natural attenuation; bioaugmentation and biostimulation (Stallwood et al. 2005). Environmental conditions determine which bioremediation approach, or combination, is most appropriate. Bioaugmentation, the addition of specialized microorganisms to enhance the biodegradation efficiency of contaminants in soil, has proved to be a feasible and economic method compared with other treatment techniques, such as chemical or physical ones, and has been received increasing attention in recent years (Vogel 1996). Monitored natural attenuation is considered the simplest bioremediation approach and comprises checking the intrinsic degradation of contaminants in an environment. Contaminated sites that have remained unchanged for long periods of time represent situations not suited for monitored natural attenuation because there is no evidence that once monitoring of the site starts, the contamination level would decrease without anthropogenic intervention. Biostimulation addresses the deficiencies of the environment, providing the optimal conditions for microbial growth, activity and thus biodegradation. Common biostimulation processes include supplementation with necessary or additional nutrients, water or air. More site-specific treatments may include chelating agents to detoxify metals or surfactants to increase hydrocarbon bioavailability. The application of biostimulants to a contaminated site can be an important factor that should account for the environmental conditions and other biodegradation limitations of the system.

Biodegradation efficiency and extent of PAHs in contaminated sites depend upon a multitude of factors, including bioavailability of the PAH to microorganisms, the characteristics of the environment (pH, temperature, concentration and availability of suitable electron acceptors), the extent of sorption of the PAH, identity of the sorbent, mass transfer rates, and microbial factors (buildup of toxic intermediates, kinetics, or the presence or absence of organisms capable of degrading a given compound). The bioavailability of PAHs is of paramount importance, which is a compound-specific phenomenon, as sorption, desorption. Physical-chemical factors affecting bioavailability include sorption, nonaqueous phase identity, the length of time the PAH has been in contact with the soil, and microbial factors. These factors are interrelated, complex, and compound-specific. The sum of these processes determine the bioavailability of a PAH to an organism. The type of soil to which PAHs are bound can have a profound influence upon the bioavailability of PAHs.

Four steps can be put forward during the biodegradation of soil-sorbed PAH. Firstly, the PAH must be desorbed from the soil into particle-associated pore water. Secondly, the dissolved PAH must travel through the pore water to the bulk aqueous fluid. Thirdly, bulk fluid movement must transport the PAH to the microbial cell. Lastly, the microbial cell must transport the PAH molecule across its cell membranes to the cell interior where metabolism occurs. Each of these steps is subject to equilibrium and kinetic effects, and each is capable of being the rate-limiting step for a particular PAHs contaminated sites.

Mass transfer between the site where a PAH is sorbed and the microbial cell is a prerequisite to biodegradation. Both mass transfer factors and biological factors are relevant. Mass transfer concerns include desorption of the PAH molecule and arrival of the PAH and the organism at the same location, either by mixing, PAH diffusion, or cell locomotion. The cell must possess the necessary enzymes to degrade the PAH, and those enzymes must be expressed at a sufficient level of activity.

The maximum potential rate of PAH biodegradation is determined by the rate of mass transfer of the PAH to the microbial cell and the intrinsic metabolic activity of the cell. The bioavailability of a compound reflects the balance between these two processes. Inherent properties such as the octanol:water partitioning coefficient (Kow) and molar volume affect the behavior of a PAH in processes such as sorption/desorption and diffusion. The hydrophobic nature of PAHs results in their partitioning onto soil matrix and their strong adsorption onto the soil organic matter (Manilal and Alexander 1991; Weissenfiels *et al*. 1992). Because of their hydrophobicity and low solubility, PAHs in a soil or sediment matrix will tend to associate with the organic matter coating of the particle matrix rather than the hydrated mineral surface. The sorption of organic contaminants by natural organic matter often limits the bioavailability of substrates, and is an important factor affecting microbial degradation rates in soils and sediments (Grosser et al. 2000). Sorption behavior of PAHs is complicated by the physical complexity and heterogeneity of the soil matrix (Luthy et al. 1997; Stokes et al. 2005). It is generally believed that adsorbed hydrophobic pollutants are not directly available to the microbial population. A well-designed bioremediation process should consider methods to mobilize these contaminants from soil surface and make them available to the microorganism. Variable results have been shown concerning the utility of using surfactants or solvents in hydrocarbon solubilization and biodegradation. It is to be expected that some of these flushing agents will remain in the treated soil after the flushing process. The potential impact of residual flushing agents on microbial processes is a question of concern.

Soil is thought to contain the greatest biodiversity of any environment on Earth. Different investigation strategies into the biodiversity in soil exist that involve of environmental sampling and extraction of target molecules and include analysis of key biogenic molecules like membrane lipid and/ or respiratory quinone profiles. The most commonly used method for microbial classification is by sequencing the 16S rRNA gene. The 16S rRNA gene provides a highly conserved marker, with a slow and constant mutational rate that can be used to measure taxonomic distances between species based on differences in the DNA sequence. Many molecular phylogenetic environmental studies using 16S gene analyses have uncovered numerous, potentially new microbial species, genera and even domains lurking, with no cultured, laboratory strain representative for comparison. Speculation of the order of magnitude concerning the total number of bacterial species is debated by microbiologists (Hong et al. 2006), making it impossible to precisely quantify the significance of the cultured laboratory stains, which may only represent ~ 1 % of the total number of species on the planet (Amann et al. 1995).

16S rRNA gene analysis introduces biases and limits the practicality of basing community profiles solely on DNA isolation, amplification and sequencing. The process can be divided into three major stages, each of which can introduce bias; DNA extraction; polymerase chain reaction (PCR); DNA sequencing and bioinformatic analyses. Various chemical and mechanical techniques exist that are designed to extract DNA from within cells and the surrounding physical matrix, and purify this separated DNA (Sambrook and Russell 2001). The efficiency for DNA extraction depends on the methods used, the physical matrix, and the cell type (Greer 2005). Although extraction methods are designed to deal with distinct matrices and cell types, for instance Gram-positive cells are generally more resistant than Gram-negative cells to lysis, no method is considered infallible (Krsek and Wellington 1999; Martin-Laurent et al. 2001).

2. Background of the contaminated site

The Beijing Coking and Chemistry Plant was constructed in 1959, and was totally suspended in July 2006 and relocated to Tangshan City, Hebei Province. The plant was located in the east of Beijing city, was adjacent to Beijing Dye Processing Plant in the west, 5th Ring Road in the east, Huagong Road in the south, old Boluoying Village in the north and Huagong Bridge in the southeast corner, as the starting point of another Jing-jin-tang Expressway and also a major part of Beijing (Fig.1).

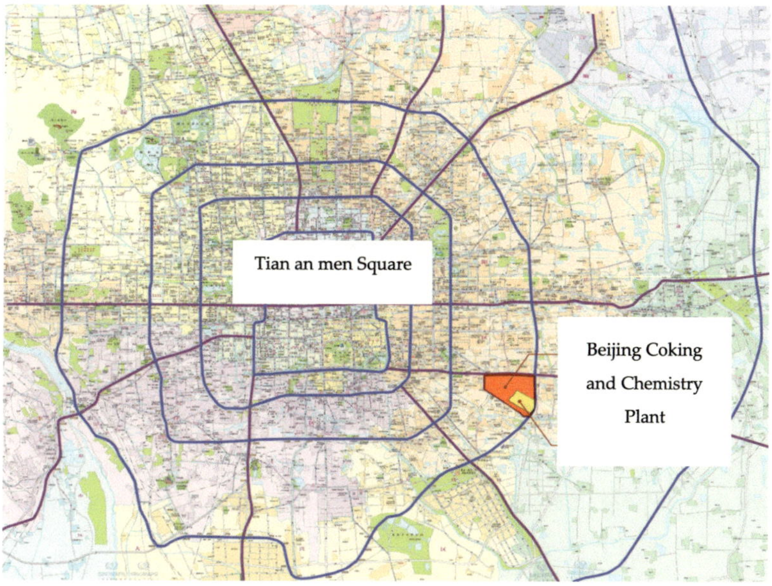

Figure 1. Location of Beijing Coking and Chemistry Plant

The identification of appropriate remediation technologies will be conditioned by knowledge of the soil conditions that exist, and of water movement and contaminant transport in the unsaturated zone. Field investigations are conducted to initially determine fundamental soil and contaminant properties, and to pilot test and evaluate remediation technologies. In addition to standard sampling at various depths, and laboratory determinations of soil organic carbon, total P, total N, total K etc., it is intended to make use of more sophisticated geophysical techniques to map shallow water tables and contamination distribution. All of the results were shown in Table1 and Table 2.

Parameter	Value
Soil organic carbon (dry weight.%)	3.32
Total P (mg P/kg soil)	0.29
Total N (mg N/kg soil)	0.81
Total K (mg K/kg soil)	14.91
Soil pH (in reagent grade water)	8.13

Table 1. The characteristics of the soil.

PAHs	Soil depth		
	0.5m	1.2m	3.0m
Acenaphthylene	11.37	10.70	7.70
Acenaphthene+ Fluorene	2.83	1.43	5.51
Anthracene	2.17	0.49	6.18
Fluoranthene	14.0	3.93	1.81
Phenanthrene	16.32	5.76	2.98
Pyrene	6.14	1.54	0.84
Benzo(a)anthracene + Chrysene	18.14	8.60	21.94
Benzo(b)fluoranthe	24.05	8.73	17.79
Benzo(k)fluoranthene	10.14	3.81	4.04
Benzo(a)pyrene	9.88	4.06	8.72
Dibenzo(a,h)anthracene	23.43	3.17	6.10
Indeno(1,2,3-cd)pyrenene	21.21	7.38	12.96
Total amount	159.68	59.59	96.57

Table 2. PAH concentration in soil of different depth (mg/kg)

3. Sampling and analytical methods

3.1. Soil and groundwater sampling and analysis

Collection of representative soil samples groundwater samples for contaminants of concern are crucial for the whole analysis process. The sampling programs would employ a judgemental sampling approach with borehole and monitoring wells being targeted to assess known areas/depths of contamination. Soil samples were taken by bores located on the site. Surface soil samples were collected from the bores and Selected distances, up to 12.74 m; in depth samples were collected at 1 m intervals. Control samples were collected from points located far away from the disposal sites, where soil contamination is practically nonexistent. The soil samples used in the experiment were collected in brown glass bottles and stored at 4°C until analysed. The samples were dried and then sieved with 2 mm diameter before experiment. The groundwater was sampled after purging the wells using a submersible pump and a Teflon hose. The samples were stored in brown glass bottles at 4°C until analyses.

Soil analysis was carried out using standard methodologies (Page, 1982). Particle size distribution was carried out using the Bouyoukos method (Bouyoucos, 1962); pH and electrical conductivity were measured in the paste extract using pH/EC meter equipped with glass electrode; Field moisture, water holding capacity (WHC)and degree of saturation were determined for each homogenized sample upon return to the laboratory. Organic matter was determined by dichromate oxidation; carbonates by using Bernard calcimeter; total N by the Kjeldahl method; available phosphorous using sodium hydrogen carbonate extraction; exchangeable K, Ca and Mg using BaCl2 extraction, while available Mn, Fe, Cu and Zn using DTPA extraction. Determination of NH4+, NO3−, Cl−, PO43−, and SO42− was performed in

1:10 water extracts using Dionex-100 Ionic Chromatography. Soil was extracted with boiling water using the azomethine-H method. Methanol extractable phenol compounds were quantified by means of the Folin–Ciocalteu colorimetric method (Box, 1983).

3.2. Microbial parameters analysis

3.2.1. Counting of the total bacterial community and PAH-degraders

The total bacteria cultured from the soil samples were quantified by mixing 10 g of soil with 50 mL of sterile Ringer solution in a Waring Blender for 1.5 min at high speed. From this soil suspension, successive 1/10 dilutions were made by adding 1 mL of the soil suspension to 9 mL of sterile Ringer's solution. An aliquot (0.1 mL) of each diluted soil suspension was spread on media constituted by Nutrient Agar (Difco, Detroit, USA). Colonies were counted after 7 d of incubation at 30°C. Total cultivated bacteria were expressed in colony-forming units per gram of dry soil (CFU g-1 of dry soil). Bacterial counts were carried out in triplicate.

Bacterial PAH-degraders were quantified in sterile polypropylene microplates (Nunc, Germany) according to Stieber et al. (1994). The wells were filled with 250μL of mineral salts medium. One of the PAHs that was most representative of the studied soil, phenanthrene or fluoranthene, was added, according to the selected test. The wells were inoculated with 25 μL of previously diluted soil suspension (from 10^{-1} to 10^{-8}). Inoculation was done in triplicate. The microplates were incubated at 30°C for 20–30 d and then were evaluated for colored products. Enumeration of bacterial PAH degraders was carried out using the Most Probable Number (MPN) method (De Man, 1977).

3.2.2. Polymerase chain reaction (PCR) amplification technique

The screening and cultivation of PAHs-degrading microorganisms usually precede the other steps in studies of PAHs microbial degradation. Once such microorganisms were obtained, identification and categorization of related microorganisms is a frequently confused technical problem. Categorization and identification of newly isolated above mentioned microorganisms traditionally depend on phenotypic characteristics such as colony and cell morphology as well as biochemical and serological characteristics such as protein and fatty acid pattern profiles. However, it is often time-consuming and highly experience-reliance. With the rapid development of molecular-biology, modern taxonomy prefers sequencing technologies of molecular markers such as 16S rRNA or 18S rRNA. These technologies allow the identification of colonies isolated from microbial consortia and the establishment of phylogenetic relationships between them (Molina et al., 2009). In addition to taxonomy, PCR combined with other approaches could also be used to estimate in situ how pollution affects the bacterial community structure and composition of sediments. Several PCR techniques such as random amplified polymorphic DNA (RAPD-PCR), arbitrarily primed-PCR (AP-PCR) could be used to identify species. Besides, bacterial 16S rRNA has become the most commonly used molecular index for its evolutionary distinctive sequence. PCR approach (16S rRNA) did make categorization and identification of interested strains convenient.

3.2.3. Fingerprinting techniques based on 16S rRNA

The applications of molecular biological techniques to detect and identify microorganisms have shown enormous wealth of microbial diversity, and at the same time the limitations of traditional cultivation techniques to retrieve this diversity. However, although successful, these studies have only focused on the exploration of microbial diversity, they have not given any information on the complex dynamics in which microbial communities can undergo by PAHs changes and seasonal fluctuations or after environmental perturbations. It is now widely accepted that it is the whole community, but not only one or several microbes that could fulfill the complex degradation of the PAHs. According to widely accepted opinion, about 99% of the microorganisms in natural environment are unculturable through traditional methods until now. In this sense, molecular technique is a preferable approach avoiding this insurmountable obstacle to discover the whole community of certain microorganisms because the DNA of majority microorganisms in certain environment could be obtained. However, the cloning approach exclusively is not efficient, because it is time-consuming and labour intensive, and hence impractical for multiple sample analysis. Fingerprinting techniques based on 16S rRNA such as DGGE (denaturing gradient gel electrophoresis) , SSCP (Single-Strand Conformation Polymor-phism), T-RFLP (terminal restriction fragment length polymorphism) and RISA (Ribosomal intergenic spacer analysis) could analysis the diversity and dynamics of the whole community at molecular level.

A single nucleotide change in a particular sequence, as seen in a double-stranded DNA, cannot be distinguished by electrophoresis, because the physical properties of the double strands are almost identical for both alleles. After denaturation, single-stranded DNA undergoes a 3-dimensional folding and may assume a unique conformational state based on its DNA sequence. The difference in shape between two single-stranded DNA strands with different sequences can cause them to migrate differently on an electrophoresis gel. SSCP used to be applicable as a diagnostic tool in molecular biology but was first introduced into environmental ecology in 1996 (Lee, 1996). The shortfall of SSCP include low reproducibility, short effective sequence (about 150–400 bp) and susceptibility of gel concentration and electrophoresis temperature. T-RFLP aims to generate a fingerprint of an unknown microbial community for profiling of microbial communities based on the position of a restriction site closest to a labeled end of an amplified gene. The major advantage of T-RFLP is the use of an automated sequencer which gives highly reproducible results for repeated samples. However, the fact that only the terminal fragments are being read means that any two distinct sequences which share a terminal restriction site will result in one peak only on the electropherogram and will be indistinguishable.

Although above mentioned several molecular methods were widely adapted in microbial ecology researches, the most commonly used method in PAHs biodegradation was DGGE. The technique is based on the electrophoretic separation of PCR-generated double-stranded DNA in an acrylamide gel containing a gradient of a denaturant. As the DNA encounters an appropriate denaturant concentration, a sequence-dependent partial separation of the double strands occurs. This conformational change in the DNA tertiary structure causes a reduced migration rate and results in a DNA band pattern representative of the sampled microbial community. The interested strands could be punched out from the DGGE gel and the enclosured DNA could be released and amplified. Thereafter, nucleotide sequence of the target DNA could be obtained and then compared with those available in GenBank to identify the closest relatives using the BLAST algorithm. On the other hand, it is widely accepted that only a pretty small part of the microorganisms community (0.1–10%) present in any environmental sample could be cultivated in general laboratory media (Amman et al. 1995) and the appliance of new developments in culture-independent methods such as DGGE/TGGE has greatly increased the understanding of more members of microbial consortia than culture-dependent approaches. Although several PAHs degrading bacterial species have been isolated, it is not expected that a single isolate would exhibit the ability to degrade completely all PAHs. A consortium composed of different microorganisms can better achieve this. PCR–DGGE of 16S rRNA gene sequences was used to monitor the bacterial population changes during PAHs degradation of the consortium when pyrene, chrysene, and benzo[a]pyrene were provided together or separately in the TLP cultures (Lafortune et al. 2009). Nevertheless, from the application point of view, cultivated microorganisms are more interested and necessary for the further extraction and purification of the enzymes involved in PAHs degradation as well as large-scale cultivation for engineering input.

3.3. Extracting DNA and enzymes from soil samples

3.3.1. DNA extractions

In the initial efforts to extract DNA from sediments and soils workers used either cell extraction (recovery of cells from the soil matrix prior to cell lysis) or direct lysis within the soil matrix (Holben et al. 1988). Direct lysis techniques, however, have been used more because they yield more DNA and presumably a less biased sample of the microbial community diversity than cell extraction techniques yield. A major drawback of direct lysis methods is that more PCR-inhibitory substances are extracted along with the DNA (Ogram et al. 1987). In addition, the number and diversity of the direct lysis DNA extraction protocols used for soils and sediments are daunting (Herrick et al. 1993), but each protocol usually includes from one to all three of the following basic elements: physical disruption, chemical lysis, and enzymatic lysis.

Four extraction procedures, method 1 to method 4 (M1–M4), were employed; all were based on the direct lysis of cells in the sample, with subsequent recovery and purification of nucleic acids. Before extraction, all solutions were rendered DNase-free by treatment with 0.1% diethyl pyrocarbonate (DEPC). Method 1 (M1) was modified from DeLong et al. (1993).

Method 1 (M1): liquid nitrogen (approx. 10 ml) was mixed with 250 mg of each biomass sample (wet weightfrom pellet) in a mortar, ground and transferred to a micro-centrifuge tube (Eppendorf, Germany), and 1 ml of cetyl trimethylammonium bromide (CTAB) extraction buffer (Griffiths et al. 2000) was added, followed by vortexing for 30 s. After the addition of 500 μl of lysis buffer (50 μM Tris–HCl [pH=8]; 40 μM ethylene diamine tetraacetic acid [EDTA; pH=8]; 750 μM filter-sterilised sucrose) and 20 μl of lysozyme (10 mg ml^{-1}; Sigma-Aldrich, Germany), mixtures were briefly vor-texed (30 s) and incubated at 37°C for 30 min. Sodium dodecyl sulphate was added to a final concentration of 2%; the samples were again vortexed and then incubated at 70°C for 1 h. After this, 6 μl of proteinase K (Sigma-Aldrich) were added. Samples were then vortexed and incubated at 50°C for a further 30 min followed by centrifugation for 15 min (10,000×g). The supernatants were transferred to fresh micro-centrifuge tubes, and the aqueous phase was extracted by mixing an equal volume of chloroform–isoamyl alcohol (24:1) followed by centrifugation (10,000×g) for 10 min.. Total nucleic acids were then precipitated from the extracted aqueous layer with 0.6 vol of isopropanol overnight, at room temperature, followed by centrifugation (10,000×g) for 15 min. The pelleted nucleic acids were washed in 70% (v/v) ice-cold ethanol and air dried before re-suspension in 50 μl DEPC-treated water.

Method 2 (M2): Soil or sediment of 250 mg and 1 ml of 1% CTAB were beaten for 2 min with 250 mg of zirconia/silica beads (1.0, 0.5 and 0.1 mm; Biospec Products, USA), in the Mini Beadbeater-8 (Biospec Products) at the median speed setting. A 500-μl aliquot of lysis buffer was added to the mixture, and the remainder of the extraction protocol was continued as described for M1.

Method 3 (M3): Briefly, 500 mg of the soil or sediment samples were added to 0.5 ml of CTAB extraction buffer and 0.5 ml of phenol–chloroform–isoamyl alcohol (25:24:1; pH 8.0) and lysed for 30 s with 250 mg of zirconia/silica beads (1.0, 0.5 and 0.1 mm), in the Mini Beadbeater-8 at the median speed setting. The aqueous phase, containing the nucleic acids, was separated by centrifugation at 10,000×g for 5 min and removed to respective fresh micro-centrifuge tubes. The aqueous phase was extracted, and phenol was removed by addition of an equal volume of chloroform–isoamyl alcohol (24:1) followed by centrifugation for 5 min (10,000×g). Two volumes of polyethelene glycol (PEG)–1.6 M NaCl (30% w/v) were used to precipitate total nucleic acids at room temperature, which were then washed with ice-cold 70% (v/v) ethanol and air dried before re-suspension in 50 μl of DEPC-treated water (Griffiths et al. 2000).

Method 4 (M4): The MoBio Ultraclean™ soil DNA kit (Cambio, Cambridge, UK). DNA was extracted from 250 mg of soil or sediment according to the manufacturer's instructions.

To inspect the quality of extracted DNA, 5-μl aliquots of crude extract were run on Tris–acetate–EDTA (TAE) agarose gels (1%) containing ethidium bromide (1 ng ml^{-1}; Maniatis et al. 1982) for DNA staining and visualisation, with Lambda DNA/HindIII molecular size marker (Promega, USA). Gel images were captured using a UV transillumination table and the AlphaDigiDoc 1201 system (Alpha Innotech, USA).

All four methods successfully extracted DNA, which was visible on agarose gels from the three soils tested. Many studies of DNA extraction techniques have reported lysis efficiencies that concur with the range observed in this study. M2 lysed most cells (93.8%), while M1 was least successful (87.7%). M3 and M4 were very similar with regards to their lysis efficiencies. M2 achieved, in almost every case, significantly higher cell lysis than the other three methods; however, there was considerable variation in the lysis efficiencies between different soil types, indicating that soil texture has a substantial impact on these measurements.

The choice of the DNA extraction method significantly influenced the bacterial community profiles generated. Greater variations were, however, observed between replicate DGGE profiles generated with M1 and M4, demonstrating that a lesser degree of reproducibility was achieved with these methods.

The results of this comprehensive evaluation of nucleic acid extraction methods suggested that M2 and M3 were both suitable for use in a large-scale study involving the direct comparative analysis of multiple soil types. The application of M2—in almost all cases—resulted in the resolution of greater diversity and employed in the research case on Beijing Coking and Chemisitry Plant.

3.3.2. *Enzymes analysis*

The microbial activities, in terms of soil enzymes, in long-term contaminated site soils were measured using the following methods. Dehydrogenase activity was determined as described by Casida et al. (1964). Soil Catalase activity was determined by the method of Cohen et al. (1970). Catalase is an iron porphyrin enzyme which catalyses very rapid decomposition of hydrogen peroxide to water and oxygen (Nelson and Cox, 2000). The enzyme is widely present in nature, which accounts for its diverse activities in soil. Catalase activity alongside with the dehydrogenase activity is used to give information on the microbial activities in soil.

Dehydrogenases (DHA) are enzymes which catalyse the removal of hydrogen atom from different metabolites (Nelson and Cox, 2000). Dehydrogenases reflect physiologically active microorganisms and thus provide correlative information on biological activities and microbial populations in soils (Rossel et al., 1997). This parameter is considered to be highly relevant to ecotoxicological testing and sensitive to PAH contaminants. DHA activity in soils was carried as following process. Six gram of soil were placed in a test tube, mixed with 0.2 g CaCO3, 1 mL TTC (2,3,5-triphenyltetrazolium chloride – as an electron acceptor) and 2.5 mL distilled water. The tubes were corked and incubated for 24hr at 37°C. The triphenylformazan (TPF) was extracted with ethanol and the intensity of the reddish colour was measured on a Beckman DU-68 spectrophotometer at a wavelength of 485 nm with ethanol as a blank. All dehydrogenases activity determinations were done in triplicate and the results were expressed as an arithmetic mean.

Soil catalase activity can be an indicator of the metabolic activities of aerobic microorganisms and correlated to the population of aerobic microorganisms and soil fertility. Catalase activity was determined by the method of Cohen where decomposed hydrogen peroxide is measured by reacting it with excess of $KMnO_4$ and residual $KMnO_4$ is measured spectrophotometrically at 480 nm. One tenth ml of the supernatant was introduced into differently labelled test tubes containing 0.5 ml of 2 mMol hydrogen peroxide and a blank containing 0.5ml of distilled water. Enzymatic reactions were initiated by adding sequentially, at the same fixed interval, 1ml of 6 N H_2SO_4 to each of the labelled test tubes containing different concentrations of spent engine oil ranging from 0.25 to 2%and to the blank sample. Also, 7 ml of 0.1N $KMnO_4$ was added within 30 s and thoroughly mixed. Spectrophotometer standard was prepared by adding 7 ml of 0.1 N $KMnO_4$ to a mixture of 5.5 ml of 0.05 N phosphate buffer, pH 7 and 1 ml of 6NH_2SO_4, the spectrophosphotometer was then zeroed with distilled water before taking absorbance readings.

3.4. Analyzing microbial diversity using PCR-DGGE

Bacterial diversity was analysed using the culture independent molecular techniques 16S rDNA gene PCR following with denaturing gradient gel electrophoresis (DGGE) . This technique proved to be a powerful tool to monitor the changes in total and PAH-degrading bacterial communities during a bioremediation process. This technique is based on the direct extraction of genomic DNA from soil samples, the amplification of 16S rRNA genes (V3 region) by using the specific primers and the separation of the PCR products by DGGE. The soils sampled in different places and different depths can be analysis in the same DGGE profiles. After the genomic DNA of samples were extracted. 16S rDNA fragments (16S rRNA gene V3 region) were amplified by using two sets of specific primers F341GC, R907 with a GC clamp in 5'end of the forward primer and F341. Initial denaturation was at 94 °C for 45 s, amplification was carried out using 30 cycles including denaturation at 94°C for 45 s, annealing at 55°C for 45 s and DNA extension at 72°C for 60 s, and final extension at 72°C for 6 min. An aliquot of 5μL of the PCR product was run in 1.5% agarose gel at 120 V for 45 min. DGGE was performed using a D-Code 16/16-cm gel system with a 1-mm gel width (Bio-Rad, Hercules, Calif.) maintained at a constant temperature of 60°C in 7L of 1xTAE buffer. The acrylamide concentration in the gel was 6% and gradients were formed between 35% and 65% denaturant. Gels were run at a constant voltage of 110 V for 11 h, at 60°C. The gels were then stained with 0.5 mg L-1 ethidium bromide solution for 30 min, destained in 1xTAE buffer for 15 min, and photographed. Pieces of the DGGE bands of total community DNA to be sequenced were cut out with a sterile scalpel and placed in Eppendorf tubes containing 50μL of sterilized milli-Q water. The DNA of each band was allowed to diffuse into the water at 4°C overnight. After centrifugation at 10,000g for 6 min, 10μL of the solution was used as the DNA template in a PCR reaction using the same conditions. PCR products were analyzed using DGGE to confirm that the expected products were isolated.

The PCR products are separated by DGGE respectively. DGGE patterns from environmental samples were compared with the migration of reference clones of known sequence, and the major environmental bands were excised, reamplified, and sequenced to investigate their identities further. Similar bands were matched after normalising the bands using Gaussian modelling and background subtraction. The matched band profiles were converted into binary data using Quantity One Software. The similarity matrix was calculated based on the Dice coefficient (Dice, 1945). The clustering algorithm of unweighted pair-group method with arithmetic averages (UPGMA) was used to calculate the dendrogram.

4. Results and discussions

4.1. Relationship between microbial population and soil contamination levels

The reduction of PAH content was linked to the bacterial degrading activity of the soil, which was kept at an optimal level with respect to humidity and oxygenation during the treatment. Soils highly contaminated by hydrocarbons displayed different microbiological properties. In particular the higher/the lower the pollutant content, the smaller/the greater are the activities of some enzymes related to nutrient cycling and the viable bacterial cell numbers. The different microbiological properties of the soils probably reflect the different bacterial diversity as assessed. Studies showed that: PAH induce perturbations in the microbial communities in terms of density and metabolism; indigenous bacteria seem to have a high capacity of PAH degradation, depending on the physicochemical properties and the availability of substances present.

A number of bacterial species are known to degrade PAHs and most of them are isolated from contaminated soil or sediments. *Pseudomonas aeruginosa, Pseudomons fluoresens, Mycobacterium* spp., *Haemophilus* spp., *Rhodococcus* spp., *Paenibacillus* spp. are some of the commonly studied PAH-degrading bacteria. *Lignolytic* fungi too have the property of PAH degradation. *Phanerochaete chrysosporium, Bjerkandera adusta,* and *Pleurotus ostreatus* are the common PAH-degrading fungi. Among the PAH in petrochemical waste, 5 and more rings PAHs always have toxicity and harder to be degraded. Fungi have prevalent capability of degrading five-ring PAHs because of their recalcitrant and hydrophobic nature (Cerniglia, 1993). Both lignin-degrading white-rot and some non-white-rot fungi (*Aspergillus niger, Penicillium glabrum, zygomycete Cunninghamella elegans, basidiomycete Crinipellis stipitaria*) have been shown to degrade a variety of pollutants including PAH, DDT, PCP, and TNT (Haemmerli et al., 1986). Benzo(a)pyrene is considered as the most carcinogenic and toxic. And some kinds of bacteria are also found having the ability to degrade BaP. Ye et al. *Sphingomonas paucimobilis* strain EPA 505 can degrade BaP and also indicated that has the significant ability of hydroxylation and ring cleavage of the 7,8,9,10-benzo ring. Aitken et al. (1991) isolated 11 strains from a variety of contaminated sites (oil, motor oil, wood treatment, and refinery) with the ability to degrade BaP. The organismswere identified as at least three species of Pseudomonas, as well as *Agrobacterium, Bacillus, Burkholderia* and *Sphingomonas* species. and BaP has been reported to be degraded by other bacteria including *Rhodococcus* sp., *Mycobacterium,* and mixed culture of *Pseudomonas* and *Flavobacterium* species (Walter et al.1991). Small molecular weight PAHs such as 2-, 3- and 4-ring PAHs are considered to be degraded easier than 5 and more rings. Christine et al (2010) used by DGGE profiles of the 16S rDNA genes assessed the bacterial community in soil contaminated by mostly 2-, 3- and 4-ring PAHs produced by coal tar distillation, and found persistence of a bacterial consortium represented by Gram-negative bacterial strains belonging mainly to *Gamma-proteobacteria,* and in particular to the *Pseudomonas* and *Enterobacter* genera .These strains had a strong PAH-degrading capacity that remained throughout the biotreatment. Thus, the presence of *Pseudomonas* and *Enterobacter* strains in this type of PAH-contaminated soil seems to be a good bioindicator for the potential biodegradation of 2-, 3- and 4-ring PAHs. Other species, such as *Beta-proteobacteria*, appeared over the course of time, when the PAH concentration was low enough to strongly decrease the ecotoxicity of the soil. Thus, the *Beta-proteobacteria* group could be a good indicator to estimate the end point of biotreatment of mostly 2-, 3- and 4-ring PAH-polluted soil to complement chemicalmethods.

In the study of Beijing Coking and Chemistry plant, the direct count of microbes from all 68 samples showed the microbes distribution varied in each borehole and increased with contaminates concentration. The minimum, 8.90×10^6 CFU/g, appeared at the depth of 4.10m, the maximum, 1.422×10^8 CFU/g, appeared at the depth of 6.80m. The reasons may be that the sites have been exposured to coal tar pollution for a long time, and indigenous microorganisms can survive after selected by the pressure of the coal tar-contaminated environment, that is "survival of the fittest". Microorganisms in this site survive well can be well adapted to the coal tar, suggesting that the number of indigenous microorganisms in the local underground environment may have a relationship with coal tar pollutants. There is obvious correspondence between the distribution rules of the number of microorganisms at vertical depth of the same drilling and the distribution rules of pollutant concentration. The numbers of microorganisms increase in the soil with higher concentrations of pollutants. In fact, the maintenance of an adequate microbial biomass in soil, with a high microbial activity, could be a mechanism for soil resistance to degradation factors and thus of paramount importance for ecosystem sustainability. We can find that there is a growing trend of the number of microorganisms as the concentration of PAHs increased (Fig.2) and it is consistent with the finding of the research of Duncan et al. (1997). The results of this study show that the characteristics of the distribution of microorganisms in the contaminated soil is an important biological factors affecting natural attenuation and biological degradation of organic pollutants of coal tar. The coal tar contamination on soil can enhance the adaptability of degrading microorganisms in soil and increase the number of indigenous degrading bacteria.

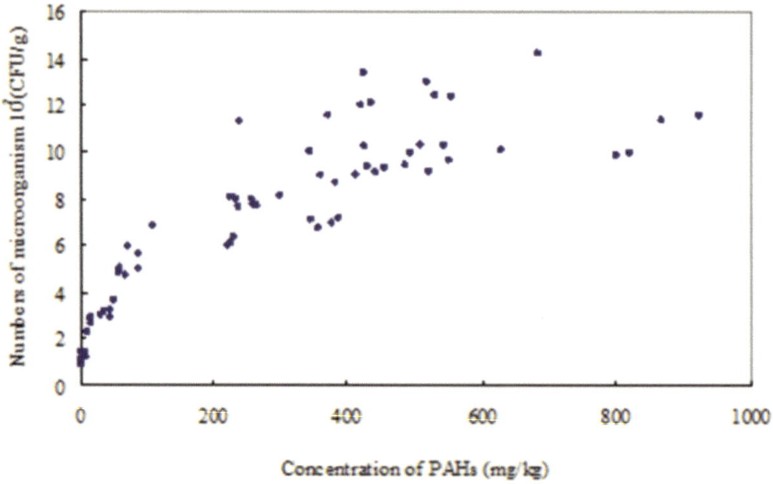

Figure 2. The relationship between the microorganisms quantities and concentrations of pollutants of sampling point

The determination and the analysis of the sequence of the advantages and the specific belt in DGGE fingerprints shows that 14 kinds of the indigenous prokaryotic microorganisms had been found in underground environment of pollution site, namely *Bacillus* sp. EPI- R1 (*Bacillus* genus EPI-R1 strains), *Bacillus* cereus.03BB102, *Bacillus thuringiensis (Bacillus thuringiensis*), *Bacillus* sp. SA, Ant14 *(Bacillus* SA Ant14 strains), the *Bacillus weihenstephanensis* KBAB4 (*Wechsler Bacillus* KBAB4), *Uncultured Bacillus* sp. (*uncultured Bacillus* certain bacteria) and *Ralstonia pickettii* (*Petri Ralston* bacteria), *Nocardioides* sp. (*Nocardia* certain bacteria), *Azoarcus* sp. BH72 (nitrogen-fixing Vibrio genus BH72 strain), *Erythrobacter litoralis* HTCC2594 (*Erythrobacter* genus HTCC2594 strain), *Magnetospirillum magneticum* (*Acidovorax Willems* genus *magnetotactic spiral* strains), *Acidovorax avenae* subsp. (food acid sub-species of the genus avenae strain), *clavibacter michiganensis* subsp. (a sub-species of *Michigan* stick rod-shaped bacteria) and an unknown new species has not been reported in GenBank.

In the research, the wildly existed bacteria underground environmental of contaminated sites can be classified as 4 groups (shows in Fig.3). (1) *Bacillus* cereus, the *Bacillus* genus, has an absolute advantage and are all found at different depth of the drilling. *Bacillus* sp. EPI-R1, the *Bacillus* genus; *Bacillus* cereus 03BB102, *Bacillus* sp.; *Bacillus thuringiensis*; *Bacillus* sp. SA, Ant14; *Bacillus weihenstephanensis* KBAB4, can also exist in various vertical depth in the venues. It means these five kinds of Bacillus genus bacteria share a common living space. However, in the *Bacillus* genus, there is a kind of bacteria named *Uncultured Bacillus* sp. different from the other five kinds. It survives in the depth range of 6.30 ~ 12.90m and is not grown in the laboratory and making the anaerobic living. (2) *Actinobacteria, Nocardioides* and *Actinobacteria Clavibacter* are the dominant *Actinobateria* to exist in this contaminated environment. *Nocardioides* sp. and *Clavibacter michiganensis* subsp are widely distributed at the vertical depth from the surface to the deepest. It indicates that these two actinomyces have the highly viability and can live in both aerobic and anaerobic environments. (3) *Alphaproteobacteria, Magnetospirillum* and *Alphaproteobacteria, Erythrobacter* are the dominant bacteria existing in the depth range of 12.20 ~ 12.70 m. The *Erythrobacter litoralis* HTCC2594 and *Magnetospirillum magneticum,* exist in the aqueous medium near the groundwater level, indicating that these two bacteria of the *Alphaproteobacteria* live an anaerobic life. (4)*Azoarcus* sp. BH72, *Ralstonia pickettii* and *Acidovorax avenae* subsp. are widely exsist in the whole of the vertical distribution of the contaminated sites, indicating that these three *Betaproteobacteria* (*Beta Proteobacteria*) bacteria can survive both in aerobic and anaerobic conditions and are facultativebacteria.

4.2. Relationship between catalase and dehydrogenase distribution and hydrocarbon contamination levels

The changes of soil catalase activities influenced by PAHs contamination are showed in Fig.4 from the study of Zhang et al. (2009). The soil catalase activities were stimulated at first and then inhibited slightly during PAHs injection within the range ~ 0.05mL (0.1M $KMnO_4$)/g (soil). Lin et al. (2005) also revealed that soil catalase activities were declined with the increase of oil concentration. At the beginning of the experiment the PAHs concentrations were low which could be as carbon sources to the indigenous microorganisms, the catalase activities were higher than the original soil before day 30. With the accumulation of PAHs in the soil, its toxicity became significant and inhibited the growth of aerobic microorganisms, the catalase activities declined after 60 days, lower than those in the unpolluted soil. This is also indicated by the environmental conditions change from aerobic to anaerobic with the continuous input of PAHs contaminants.

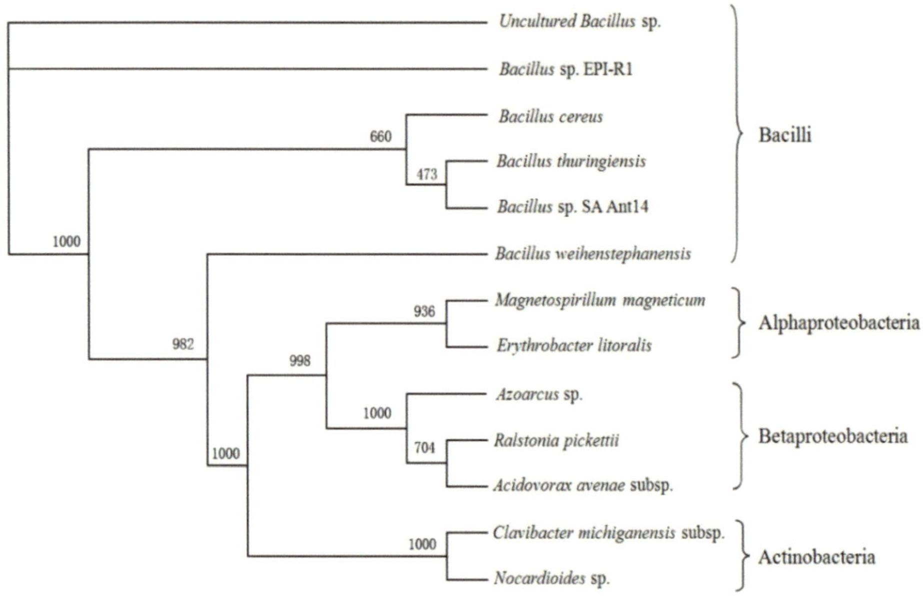

Figure 3. The 16S rDNA phylogenetic tree of dominant prokaryotic microorganisms in underground environmental of contaminated sites

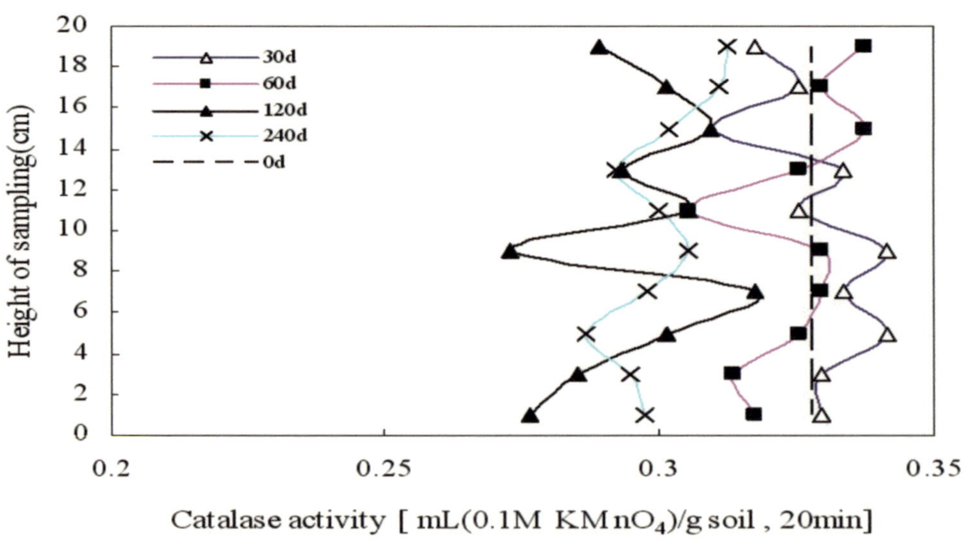

Figure 4. Change of soil catalase activities

In the present study, it was clearly shown that a positive correlation between dehydrogenase activity and the content of phenanthrene and also fluoranthene, chrysene and dibenzo[ah]anthraacene was observed. We can conclude that the effect of PAHs contamination on soil dehydrogenase activity. There was a progressive increase in the values obtained as the concentrations of the PAHs contamination increased and this was found to be significant between in soil contamination relative to the control value. Also dehydrogenase activity depends not only on the PAH type, but also primarily on the other compounds from this group and also their amount.

In the case of Beijing Coking and Chemistry plant, the spatial trends of the catalase activity at different depths of vertical drilling on the whole are the same. For each drilling, the catalase activity of the surface is higher and the catalase activity of the deep drillings is lower, the overall trend is that the catalase activity decreases with depth, indicating that the distribution characteristics of aerobic microorganisms. Catalase activity of the contaminated sites of all 68 sampling points range between 0.554 mL 0.02M $KMnO_4$ /g•h~6.230 mL 0.02M $KMnO_4$ /g•h; while the difference of catalase activity in the same formation of the entire site is not particularly large.

As to dehydrogenase in this case, the result of measurement show that the trend of dehydrogenase activity overall is the same at different depths drilling vertically, the range of the change is between 0.14 μg TF/g•h~2.54 μg TF /g•h. The differences of dehydrogenase activity are not particularly large, the reason maybe that the indigenous microorganisms of the different vertical depth of the contaminated sites can generate dehydrogenase and to some extent all have the ability to degrade organic pollutants of coal tar.

Distribution rules of dehydrogenase activity in vertical depth of the same drilling and the rules of the distribution of number of microorganisms have a certain correlation. Fig.5 shows the relationship between the dehydrogenase activities with numbers of microorganisms in this sampling site. Activities of dehydrogenase can be enhanced in the soil with more numbers of microorganisms. It means that the selection of coal tar contamination on soil microorganisms can enhance the adaptability of microorganisms and the ability of degradation.

In the vertical spatial scale, the distribution rules of the concentrations of pollutants in the site and the distribution of number of microorganisms and dehydrogenase activity has a certain correlation. The higher concentrations of pollutants in the soil, the more numbers of microorganisms and the greater the activities of dehydrogenase produced by indigenous microorganisms with of the degradation function.

4.3. Relationship of microbial diversity with soil contamination levels

PAHs present in soil may exhibit a toxic activity towards different plants, microorganisms and invertebrates. Microorganisms, being in intimate contact with the soil environment, are considered to be the best indicators of soil pollution. In general, they are very sensitive to low concentrations of contaminants and rapidly response to soil perturbation. An alteration of their activity and diversity may result, and in turn it will reflect in a reduced soil quality (Schloter et al., 2003). Hydrocarbon contamination may affect soil microbial structure. However, in soil microbiology there is a lack of knowledge regarding the environmental determinants of microbial population variation in soil environments contaminated by complex hydrocarbon mixtures (Hamamura et al., 2006). In this sense, different authors have observed, by denaturing gradient gel electrophoresis (DGGE) or terminal restriction fragment length polymorphism (T-RFLP), a decrease in the microbial diversity of fuel contaminated soils, leading to the predominance of well-adapted microorganisms (Ahn et al., 2006), and a change in the community structure (Labb′e et al., 2007).

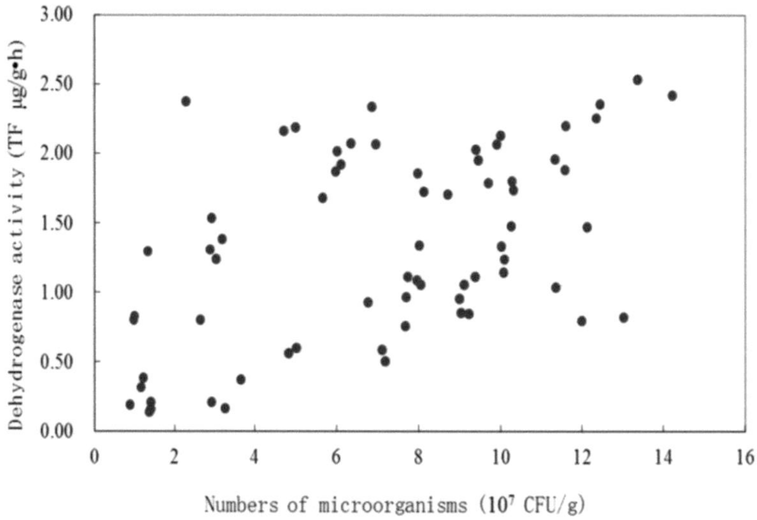

Figure 5. The relationship between the dehydrogenase activities with microorganisms quantities in the sampling sites

When bacterial diversity is analyzed by DGGE profiles of the 16S rDNA genes, the number of DGGE bands was taken as an indication of species in each sample. The relative surface intensity of each DGGE band and the sum of all the surfaces for all bands in a sample were used to estimate species abundance (Fromin et al., 2002; Sekiguchi et al., 2002). The different microbiological properties of the soils reflect the different bacterial diversity as assessed by DGGE profiles of the 16S rDNA genes.

The Shannon index, sometimes referred to as the Shannon-Wiener index or the Shannon-Weaver index, was used to evaluate the biodiversity of both soils and enrichment cultures. The Shannon-Weaver diversity index is a general diversity index which increases with the number of species and which is higher when the mass is distributed more evenly over the species. The evenness is independent of the number of species. Evenness is lower if a small number of bands are dominant and highest if the relative abundance of all bands is the same. The equitability correspondingly indicates whether there are dominant bands. The Shannon index of soils was calculated on the basis of the number and intensity of bands present on DGGE samples, run on the same gel, as follows: $H=-\sum P_i \log P_i$, where P_i is the importance probability of the bands in a gel lane. P_i was calculated as follows: $P_i = n_i/N$, where n_i is the band intensity for each individual band and N is the sum of intensities of bands in a lane. *H* was used as a parameter that reflects the structural diversity of the dominant microbial community.

In particular the lower / the higher the parameter *H*, the higher/ the lower the pollutant content, the smaller/the greater are the activities of some enzymes related to nutrient cycling and the viable bacterial cell numbers. (Andreoni et al., 2004).

In the study of Beijing Coking and Chemistry plant, the relationship of microbial diversity with soil contamination levels was show in Fig.6. Generally there are more numbers of the microorganisms in contaminated soil. However the result shows the relationship between the microbial diversity of different depth of the drilling and the soil contamination levels is not clear. The trend of the microbial diversity is also not clear. It means that the microbial diversity not only decides by the contamination levels and it is the result of the coaction of various environmental factors. The relational graph of the relationship between the prokaryotic microbial community diversity and concentration of pollutants of 68 sampling points of the underground environment also shows that the correlation between the prokaryotic microbial diversity and concentration of pollutants is not obvious.

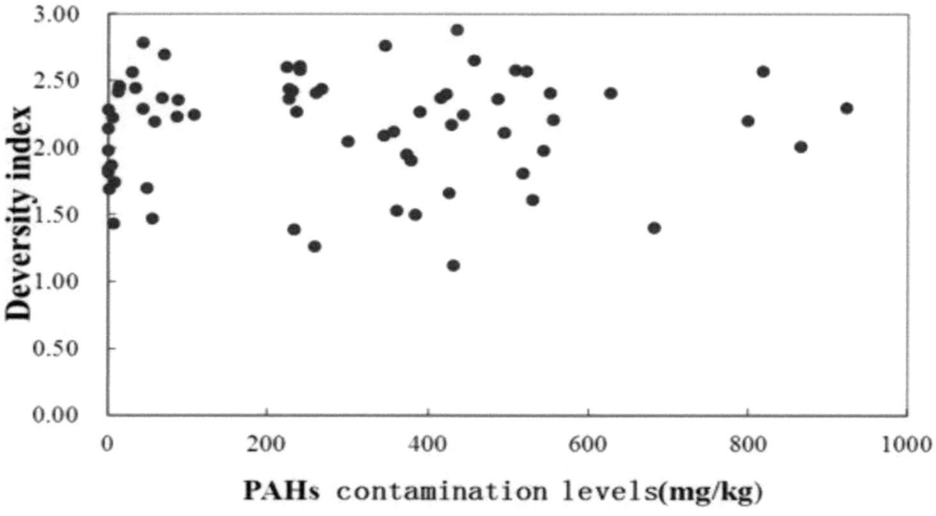

Figure 6. The relationship between the prokaryotic microbial diversity and concentration of pollutants

5. Conclusions

In the case of Beijing Coking and Chemistry plant, dehydrogenase and catalase activities were analysis to assess soil biological activity. Dehydrogenase activity (DHA), a very sensitive soil enzyme to pollutants, showed significant variation between soils. The research showed that the trend of dehydrogenase activity overall is the same at different depths drilling vertically, the range of the change is between 0.14 μg TF / g • h ~ 2.54 μg TF / g • h.

The higher concentrations of pollutants in the soil, the more numbers of microorganisms and the greater the activities of dehydrogenase activity produced by indigenous microorganisms with the degradation function. As to catalase, for each drilling, the catalase activity of the surface is higher and the catalase activity of the deep drillings is lower, the overall trend is that the catalase activities decreased with depth, indicating that the distribution characteristics of aerobic microorganisms.

The sequencing analysis of the specific belts in DGGE fingerprints showed that the prokaryotic microbial community structure had a certain degree of changes with vertical depth of the site and the DGGE finger-print of the six holes were so different. The analysis of the sequence of the advantages and the specific bands in DGGE fingerprints showed that 14 kinds of the indigenous prokaryotic microorganisms had been found in underground environment of pollution site. They could be classified into 4 classes by Phylogenetic analysis: *Bacilli, Alphaproteobacteria, Betaproteobacteria, Actinobacteria.*

Author details

Aizhong Ding, Yujiao Sun, Junfeng Dou and Xiaohui Zhao
College of Water Sciences, Beijing Normal University, China

Lirong Cheng, Lin Jiang and Dan Zhang
Beijing Municipal Research Institute of Environmental Protection, China

Acknowledgement

This work was supported by the National Natural Science Foundation of China (No. 51008026, 40873076 and 41011130204).

6. References

Ahn, J.H., Kim, M.S., Kim, M.C., Lim, J.S., Lee, G.T., Yun, J.K., Kim, T., Kim, T. and Ka, J.O. (2006). Analysis of bacterial diversity and community structure in forest soils contaminated with fuel hydrocarbon. *J. Microbiol.Biotechn,* Vol.16, No.5, (2006), pp. 704-715, ISSN 1017-7825

Aitke, M.D., Stringfellow, W.T., Nage, R.D.l, Kazung, C.a, Chen, S.H. (1998). Characteristics of phenanthrene-degrading bacteria isolated from soils contaminated with polycyclic aromatic hydrocarbons. *Can. J. Microbiol,* Vol.44, No.8, (1998), pp. 743–752, ISSN 0008-4166

Amann, R.l, Ludwig W., Schleifer, K.H. (1995). Phylogenetic identification and in situ detection of individual microbial cells without cultivation. *Microbiol,* Vol.59, No.1, (1995), pp. 143-169, ISSN 0001-3714

Andreoni, V., Cavalca, L., Rao, M.A., Nocerino, G., Bernasconi, S., Dell'Amico, E., Colombo, M., Gianfreda, L. (2004). Bacterial communities and enzyme activities of PAHs polluted soils. *Chemosphere,* Vol.57, No.5, (2004), pp. 401–412, ISSN 0045-6535

Boonchan, S., Britz, M.L., Stanley, G.A. (2000). Degradation and mineralisation of high-molecular weight polycyclic aromatic hydrocarbons by defined fungal-bacterial cocultures. *Appl. Environ. Microbiol*, Vol.66, No.3, (2000), pp. 1007–1019, ISSN 0099-2240

Bouyoucos, G.J. (1962). Hydrometer method improved for making particle and size analysis of soils. *Agron. J.*, Vol.54, No.5, (1962), pp. 464–465, ISSN 0002-1962

Box, J.D. (1983). Investigation of the Folin–Ciocalteu phenol reagent for the determination of polyphenolic substances in natural waters. *Water Res.*, Vol.17, No.5, (1983), pp. 511–525, ISSN 0043-1354

Casida, L.E., Klein, D.A. and Santoro, T. (1964). Soil dehydrogenase activity. *Soil Sci.*, Vol.98, No.6, (December 1964), pp. 371–376, ISSN 0038-0768

Cerniglia, C.E. (1993). Biodegradation of polycyclic aromatic hydrocarbons. *Biodegradation*, Vol.3, No.2-3, (1993), pp. 227–244, ISSN 0923-9820

Christine, L., Annemie, R., Frédéric, P. (2010). Evolution of bacterial community during bioremediation of PAHs in a coal tar contaminated soil. *Chemosphere*, Vol.81, No.10, (2010), pp. 1263-1271, ISSN 0045-6535

Cohen G.J., Dembiec D. and Marcus J. (1970). Measurement of catalase activity in tissue extracts. *Anal. Biochem.*, Vol.34, (Mar 1970), pp. 30-38, ISSN 0003-2697

Dan, Z., Aizhong, D., Wenjie, T., Cheng, C., Lirong, C., Yujiao, S., Xueyu, L. (2009). Microcosm study on response of soil microorganisms to PAHs contamination. *3rd International Conference on Bioinformatics and Biomedical Engineering*, ISBN, Beijing, 2009

De Man, J.C. (1971). MPN tables for more than one test. *Eur. J. Appl. Microbiol.*, Vol.4, No.4, (1971), pp. 307-316, ISSN 0175-7598

DeLong, E.F., Franks, D.G., Allredge, A.D.L. (1993). Phylogenetic diversity of aggregate-attached vs. free-living marine bacterial assemblages. *Limnol Oceanogr*, Vol.38, No.5, (1993), pp. 924-934, ISSN 0024-3590

Dice, L.R. (1945). Measures of the amount of ecologic association between species. *Ecology*, Vol.26, No.3, (1945), pp. 297-302, ISSN 0012-9658

Fromin, N., Hamelin, J., Tarnawski, S., Roesti, D., Jourdain Miserez, K., Forestier, N., Teyssier-Cuvelle, S., Aragno, M., Rossi, P. (2002). Statistical analysis of denaturing gel electrophoresis (DGE) fingerprinting patterns. *Environ. Microbiol.*, Vol.4, No.11, (2002), pp. 634-643, ISSN 1420-2026

Griffiths, R.I., Whitely, A.S., O'Donnell, A.G., Bailey, M.J. (2000). Rapid method for Co-extraction of DNA and RNA from natural environments for analysis of ribosomal DNA- and rRNA-based microbial community composition. *Appl. Environ. Microbiol.*, Vol.66, No.12, (2000), pp. 5488-5491, ISSN 0099-2240

Grosser, R.J., Friedrich, M., Ward, D.M. and Inskeep, W.P. (2000). Effect of model sorptive phases on phenanthrene biodegradation: different enrichment conditions influence bioavailability and selection of phenanthrene-degrading isolates. *Appl. Environ. Microbiol.*, Vol.66, No.7, (2000), pp. 2695-2702, ISSN 0099-2240

Haemmerli, S.D., Leisola, M.S., Sanglard, D., Fiechter, A. (1986). Oxidation of benzo(a)pyrene by extracellular ligninases of Phanerochaete chrysosporium. Veratryl alcohol and stability of ligninase. *J. Biol. Chem.*, Vol.261, No.15, (1986), pp. 6900-6903, ISSN 0021-9258

Hamamura, N., Olson, S.H., Ward, D.M. and Inskeep, W.P. (2006). Microbial population dynamics associated with crude-oil biodegradation in diverse soils. *Appl. Environ. Microb.*, Vol.72, No.9, (2006), pp. 6316-6324, ISSN 0099-2240

Herrick, J.B., Madsen, E.L., Batt, C.A. and Ghiorse, W.C. (1993). Polymerase chain reaction amplification of naphthalene catabolic and 16S rRNA gene sequences from indigenous sediment bacteria. *Appl. Environ. Microbiol.*, Vol.59, No.3, (1993), pp. 687-694, ISSN 0099-2240

Holben, W.E., Jansson, J.K., Chelm, B.K. and Tiedje. J.M. (1988). DNA probe method for the detection of specific microorganisms in the soil bacterial community. *Appl. Environ. Microbiol.*, Vol.54, No.3, (1988), pp. 703-711, ISSN 0099-2240

Hong, S.H., Bunge, J., Jeon, S.O., Epstein, S.S. (2006). Predicting microbial species richness. *PNAS*, Vol.103, No.1, (2006), pp. 117-122, ISSN 0027-8424

Islam, K.R., Weil, R.R. (1998). Microwave irradiation of soil for routine measurement of microbial biomass carbon. *Biology and Fertility of Soils*, Vol.27, No.4, (1998), pp. 408-416, ISSN 0178-2762

Johnson, A.C., Larsen, P.F., Gadbois, D.F. and Humason, A.W. (1985). The distribution of polycyclic aromatic hydrocarbons in the surficial sediments of Penobscot Bay (Maine,USA) in relation to possible sources and to other sites worldwide. *Marine Environmental Research*, Vol.15, (1985), pp. 1-16, ISSN 0141-1136

Krsek, M., Wellington, E.M.H. (1999). Comparison of different methods for the isolation and purification of total community DNA from soil. *Journal of Microbiological Methods*, Vol.39, No.1, (1999), pp. 1-16, ISSN 0167-7012

Labb`e, D., Margesin, R., Schinner, F., Whitten, L.G. and Greer, C.W. (2007). Comparative phylogenetic analysis of microbial communities in pristine and hydrocarbon-contaminated Alpine soils. *FEMS Microbiol. Ecol.*, Vol.59, No.2, (2007), pp. 466-175, ISSN 0168-6496

Lafortune, I., Juteau, P., Déziel, E., Lépine, F., Beaudet, R., Villemur, R. (2009). Bacterial diversity of a consortium degrading high-molecular-weight polycyclic aromatic hydrocarbons in a two-liquid phase biosystem. *Microb. Ecol.*, Vol.57, No.3, (2009), pp. 455-468, ISSN 0095-3628

Lee, D.H., Zo, Y.G, Kim, S.J. (1996). Nonradioactive methods to study genetic profiles of natural bacterial communities by PCR-single strand conformation polymorphism. *Appl. Environ. Microbiol.*, Vol.62, No.9, (1996), pp. 3112-3120, ISSN 2695-2702

Lim, L.H., Harrison, R.M. and Harrad, S. (1999). The contribution of traffic to atmospheric concentrations of polycyclic aromatic hydrocarbons. *Environmental Science and Technology*, Vol.33, No.20, (1999), pp. 3538-3542, ISSN 1735-1472

Lin, Q., Mendelssohn, I.A., Henry, C.B., Roberts, P.O., Walsh, M.M. (1999). Effects of Bioremediation Agents on Oil Degradation in Mineral and Sandy Salt Marsh Sediments. *Environmental Technology*, Vol.20, No.8, (1999), pp. 825-837, ISSN 0959-3330

Lin, X., Li, J.P, Sun, T.H., Tai, P.D., Gong, Z.Q. and Zhang, H.R. (2005). Bioremediation of petroleum contaminated soil and its relationship with soil enzyme activities. *Chinese Journal of Ecology*, vol. 24, No.10, (2005), pp. 1226-1229, ISSN 1000-4890

Luthy, R.G., Aiken, G.R., Brusseau, M.L., Cunningham, S.D., Gschwend, P.M., Pignatello, J.J., Reinhard, M., Traina, S.J., Weber, J., Walter, J. and Westall, J.C. (1997). Sequestration of hydrophobic organic contaminants by geosorbents. *Environmental Science and Technology,* Vol.31, No.2, (1997), pp. 3341-3347, ISSN 1735-1472

Maniatis, T., Fritsch, E.F., Sambrook, J. (1982). *Molecular cloning: a laboratory manual,* Cold Spring Harbor Laboratory Press, Cold Spring Harbor, ISBN 0879697725, New York

Manilal, V.B. and Alexander, M. (1991). Factors affecting the microbial degradation of phenanthrene in soil. *Applied and Microbiology and Biotechnology,* Vol.35, No.3, (1997), pp. 401-405, ISSN 0175-7598

Martin-Laurent, F., Philippot, L., Hallet, S., Chaussod, R., Germon, J.C. (2001). DNA Extraction from Soils: Old Bias for New Microbial Diversity Analysis *Methods. Appl. Environ. Microbiol.,* Vol.67, No.5, (2001), pp. 2354-2359, ISSN 0099-2240

Molina, M.C., González, N., Bautist, L.F., Sanz, R., Simarro. R., Sánchez, I., Sanz, J.L., Isolation and genetic identification of PAH degrading bacteria from a microbial consortium. *Biodegradation,* Vol.20, No.6, (2009), pp. 789-800, ISSN 0923-9820

Nelson, D.L. and Cox, M.M. (2000). *Principles of Biochemistry,* Macmillan Press, ISBN 1429208929, London, UK

Ogram, A., Sayler, G.S. and Barkay, T. (1987). The extraction and purification of microbial DNA from sediments. *J. Microbiol. Methods,* Vol.39, No.11, (1987), pp. 57-66, ISSN 0167-7012

Olah, G.A., Molnar, A. (2003). *Hydrocarbon Chemistry,* Wiley-blackwell, ISBN 0470935685, London, UK

Page, A.L., Miller, R.H., Keeney, D.R. (1982). *Methods of Soil Analysis, Part 2: Chemical and Microbiological Properties,* American Society of Agronomy, ISBN 0891180850, Madison, Wisconsin, US

Ping, L., Luo, Y., Wu, L., Qian, W., Song, J., Christie, P. (2006). Phenanthrene adsorption by soils treated with humic substances under different. *Environ Geochem Health,* Vol.28, No.1-2, (2006), pp. 189-195, ISSN 0269-4042

Rogers, J.C. and Li, S. (1985). Effect of metals and other inorganic ions on soil microbial activity. Soil dehydrogenase assay as a simple toxicity test. *Bull. Environ. Contam. Toxicol.,* Vol.34, No.6, (1985), pp. 858-865, ISSN 0007-4861

Sauvé, A., Dumestre, M., McBride, J., Gillett, J., Berthelin, J., Hendershot, W. (2001). Nitrification potential in field-collected soils contaminated with Pb and Cu. *Applied Soil Ecology,* Vol.12, No.1, (2001), pp. 29-39, ISSN 0929-1393

Schloter, M., Dilly, O., Munch, J.C. (2003). Indicators for evaluating soil quality. *Agricult. Ecosyst. Environ.,* Vol.98, No.1-3, (2003), pp. 255-262, ISSN 0167-8809

Sekiguchi, H., Watanabe, M., Nakahara, T., Xu, B., Uchiyama, H. (2002). Succession of bacterial community structure along the Changjiang river determined by denaturing gradient gel electrophoresis and clone library analysis. *Appl. Environ. Microbiol.,* Vol.68, No.10, (2002), pp. 5142-5150, ISSN 0099-2240

Smith, M.R. (1990). The biodegradation of aromatic hydrocarbons by bacteria. *Biodegradation,* Vol. 1, No.2-3, (1990), pp. 191-206, ISSN 0923-9820

Stallwood, B., Shears, J., Williams, P.A., Hughes, K.A. (2005). Low temperature bioremediation of oil-contaminated soil using biostimulation and bioaugmentation with a Pseudomonas sp. from maritime Antarctica. *Journal of Applied Microbiology,* Vol.9, No.4, (2005), pp. 794-802, ISSN 1364-5072

Stieber, M., Haeseler, F., Werner, P., Frimmel, F.H. (1994). A rapid screening method for microorganisms degrading polycyclic aromatic hydrocarbons in microplates. *Appl. Microbiol. Biotechnol.*, Vol.40, No.5, (1994), pp. 753-755, ISSN 0175-7598

Stokes, J., Paton, G. and Semple, K. (2005). Behavior and assessment of bioavailability of organic contaminants in soil: relevance for risk assessment and remediation. *Soil Use and Management*, Vol.21, No.2, (2005), pp. 794-802, ISSN 1364-5072

Tabatabai, M.A., Bremner, J.M. (1972). Assay of urease activity in soils. *Soil Biology and Biochemistry*, Vol.4, No.4, (1972), pp. 479-487, ISSN 0038-0717

Vogel, T.M. (1996) Bioaugmentation as a soil bioremediation approach. *Current Opinion in Biotechnology*, Vol.7, No.3, (1996), pp. 311-316, ISSN 0958-1669

Walter, U., Beyer, M., Klei, J., Rehm, H.J. (1991). Degradation of pyrene by Rhodococcus sp. UW1. *Appl. Microbiol. Biotechnol,* Vol.34, No.5, (1991), pp. 671-676, ISSN 0175-7598

Weissenfiels, W.D., Klewer, H.J. and Langhoff, J. (1992). Adsorption of polycyclic aromatic hydrocarbons (PAHs) by soil particles: influence on biodegradability and biotoxicity. *Applied Microbiology and Biotechnology,* Vol.36, No.5, (1992), pp. 689-696, ISSN 0175-7598

Chapter 8

Extracellular Electron Transfer in *in situ* Petroleum Hydrocarbon Bioremediation

Kerstin E. Scherr

Additional information is available at the end of the chapter

http://dx.doi.org/10.5772/53290

1. Introduction

1.1. Environmental pollution with petroleum hydrocarbons

Anthropogenic contamination of soil, subsurface and surface waters and atmosphere with toxic organic chemicals is an environmental issue of world wide concern. While energy and goods production from fossil hydrocarbon sources is one of the driving factors of global economy, the often adverse environmental effects of exploration, production, transport and processing of crude and shale oils, tar sands, coal and natural gas seldom come into the focus of attention.

Chemically, the term *hydrocarbons* encompasses a large variety of compounds. Saturated and unsaturated structures composed of hydrogen and carbon make up the largest fraction of organic compounds derived from fossil sources, including crude oil, its derivatives and distillates as well tar oils originating from coal gasification. These are often referred to as petroleum hydrocarbons (Cermak, 2010), also encompassing mono- and polycyclic aromatic structures of environmental and toxicological concern (Loibner et al., 2004).

In the US alone, between 5,000 and 7,000 spills of hazardous material occur every year (Scholz et al., 1998). Although the US-EPA's *Spill Prevention Control and Countermeasure Program* for inland oil spills has reduced the amount of undesired hydrocarbon release to less than 1% of the total volume of oil handled each year (US-EPA, 2012), considerable surface areas and contaminated ground water bodies remain to be decontaminated. Microbial, physical and chemical remediation strategies are available.

1.2. *In situ* bioremediation of petroleum hydrocarbons

For soil and water bodies contaminated with organic pollutants, *bioremediation* refers to the engineered exploitation of an ecosystem's intrinsic capacity for the attenuation of adverse biogeochemical influences. In *in situ* remediation soil, water and contamination remain in place[1]. These processes are driven by a large variety of microorganisms, including bacteria, fungi and archaea (*microbial bioremediation*) and may also involve plants (*phytoremediation*). The targeted addition of enriched or modified microorganisms (*bioaugmentation*) is justified in case the resident microbial population is incapable or incapacitated to perform substantial contaminant transformation or, more importantly, mineralisation. In contrast, the naturally present microbial population at a long-term contaminated site is expected to have the potential metabolic capacity for the degradation of aromatic and aliphatic petroleum hydrocarbons under both aerobic and anaerobic conditions (Widdel and Rabus, 2001). In *biostimulation* efforts, growth factors are administered to the contaminated area, including electron acceptors or donors for biochemical contaminant oxidation or reduction, respectively, and / or of micro-and macronutrients, vitamins and trace elements and others (Bartha, 1986; Cerniglia, 1992). A variety of strategies is in application or under development.

Implementing *in situ* bioremediation is an intricate task for designing and executive engineers, while decreasing the burden to economy and environment in lessening the need for soil excavation. Some strategies of physical, chemical and also microbial remediation involve harsh and cost-intensive measures effecting substantial changes to the biogeochemical conditions in the treated environments. Thus, the optimization of existing strategies and the development of novel approaches to deal with environmental pollution in sustainable ways is required both from economic and environmental viewpoints.

1.3. Objective

This review is dedicated to elucidating mechanisms for extracellular electron transfer to geogenic electron acceptors, which can potentially be coupled to oxidative hydrocarbon detoxification. An increase in the accessibility of naturally present electron accepting compounds to hydrocarbon-degrading bacteria could substantially improve bioremediation efficacy of anaerobic petroleum hydrocarbon contaminated aquifers.

2. Petroleum hydrocarbon degradation under different terminal electron accepting conditions

Petroleum hydrocarbons such as normal, branched and cycloalkanes and aromatic hydrocarbons are degradable *in situ* via biochemical oxidation both under aerobic and anaerobic conditions (DeLaune et al., 1980), provided the degradative activity is not inhibited by a lack of nutrients (nitrogen, phosphor, potassium), electron acceptors, trace elements and moisture, of if pH, temperature, salinity and contaminant concentrations are outside certain boundaries (Bartha, 1986; DeLaune et al., 1980). In contrast, hydrocarbons carrying strongly electron-withdrawing substituents are preferentially degraded via

[1] In contrast, *ex situ* bioremediation refers to manipulation strategies of excavated or delocalized contaminated matrices (e.g. in biopiles, biofilters and others) intended to decrease the contaminant load via microbial degradation.

contaminant reduction, such as sequential *reductive dechlorination* for chlorinated solvents (Maymo-Gatell et al., 1999).

A large range of microorganisms have the metabolic capability of oxidative hydrocarbon degradation using a variety of terminal electron acceptors (TEA) in the vadose and saturated zones of the subsurface (Heider et al., 1998).

2.1. Molecular oxygen as TEA

In the range of naturally occurring TEAs, molecular oxygen yields the largest amount of energy in terms of ATP production and is associated with high degradation rates with aerobic microbial communities. Molecular oxygen, however, has a low aqueous solubility (up to 10 mg/L). Fast oxygen depletion in microbial contaminant oxidation processes is opposed to slow diffusive replenishing in the groundwater. Thus, petroleum hydrocarbon contaminated aquifers are often depleted of molecular oxygen, while other, energetically less favourable TEA are reduced more slowly. Oxygen limitation is thus manifested in low natural contaminant attenuation rates, and anaerobic conditions are often encountered at biologically active contaminated sites.

2.2. Soluble alternative TEA

This has prompted the development of strategies for the stimulation of anaerobic hydrocarbon degradation. One approach is the addition of naturally occurring, well-soluble alternative terminal electron acceptors such as nitrate and sulphate (Hasinger & Scherr et al., 2012; Bregnard et al., 1998). Hydrocarbon oxidation coupled to nitrate reduction, however, yields approximately one tenth of the biochemical energy of aerobic degradation per mole hydrocarbon, and sulphate reduction about 1%. Both terminal electron accepting processes are associated with significantly lower degradation rates compared to aerobic degradation (Widdel and Rabus, 2001). The use of well soluble and less chemically reactive alternative electron acceptors, however, is connected to an increase in influence *radii* when introduced into the aquifer (Hasinger & Scherr et al., 2012).

Considerable amounts of alternative TEA, however, are required for aquifer remediation. On a theoretical mass basis, around 5 g of nitrate and sulphate are needed for the anaerobic mineralisation of 1 g of *n*-hexadecane, a mid-weight petroleum hydrocarbon, although lower demands were observed in practice (Hasinger & Scherr et al., 2012), are required. In addition, possible by-products of alternative TEA consumption, including products from incomplete nitrate reduction and sulphide precipitation, can render the implementation of these strategies under field conditions an intricate task.

2.3. Poorly soluble alternative TEA

Despite their poor aqueous solubility[2], and thus originally somewhat surprising, naturally abundant solid-phase minerals are known to play a quantitatively important role in

[2] *Poor aqueous solubility* and *insoluble in water* are often used as synonyms in this context.

subsurface microbial reduction and oxidation[3] processes (Weber et al., 2006). *Poorly soluble alternative TEA* refers mainly to iron (Fe) and manganese (Mn) minerals in their tri- and tetravalent oxidation states (Lovley et al., 2004). Their oxides, hydroxides and oxyhydroxides are amongst the most abundant redox-active elements on the earth's surface.

This comes surprising since solubility in water is deemed a prerequisite for chemicals to participate in microbial transformation processes. The participation of poorly soluble minerals in environmental redox-processes, however, is made possible by a variety of bacterial strategies to access and transfer electrons to (and from) poorly soluble mineral surfaces. These strategies include the secretion of chelators, production of electrically conductive protein structures and the participation of redox active natural organic moieties, or *electron shuttles*.

These processes are summarized as mechanisms for *extracellular electron transfer*, and are the subject of this review.

2.4. Potential use of extracellular electron transfer in bioremediation

Poorly soluble, naturally abundant alternative electron acceptors have the potential to be incorporated into efficient bioremediation strategies. The abundance of reducible iron and manganese minerals in the subsurface, may, provided these minerals can be involved into microbial hydrocarbon oxidation, alleviate the need to artificially introduce electron acceptors, such as molecular oxygen, nitrate and sulphate.

There is evidence that iron and manganese respiration can be coupled to anaerobic contaminant oxidation of monoaromatic compounds (Kunapuli et al., 2008; Lovley et al., 1989). Under field conditions, however, pronounced iron reduction coupled to the oxidation of higher molecular weight hydrocarbons is seldom reported (Aichberger et al., 2007). These data, however, indicate that hydrocarbon oxidation coupled to solid-state mineral respiration is possible, however at limited rates. The rates of Mn and Fe respiration, on the other hand, can be significantly increased via the enhancement of extracellular transport mechanisms. *Geobacter sulfurreducens,* a model dissimilatory metal reducing bacterium (DMRB) transfers electrons 27 times faster to a transient electron storage, an electron shuttle or mediator, than to iron hydroxide (Jiang and Kappler, 2008). The introduction of such electron transfer facilitators is one of the key issues where extracellular electron transport mechanisms could be incorporated into petroleum hydrocarbon bioremediation. Extracellular electron shuttles are reversibly reduced and oxidized in many subsequent cycles. Thus, only substoichiometric amounts would theoretically be required to achieve substantially increased degradation rates.

Provided a comprehensive understanding of the underlying mechanisms, efficient bioremediation strategies may be devised to enhance extracellular electron transfer from polluting organic substances to poorly soluble mineral surfaces, adjoined by oxidative pollutant transformation and detoxification. The mechanisms of electron shuttling and other

[3] Reduction / oxidation processes are also abbreviated as *redox*-processes.

extracellular electron transfer strategies devised by DRMBs, thermodynamic considerations and microorganisms involved are reviewed in Chapters 3 through 8.

In summary, possible strategies to enhance anaerobic petroleum hydrocarbon degradation using naturally occurring, poorly soluble TEA include the addition of electron shuttling compounds (ES) of natural or anthropogenic origin, as was demonstrated for the anaerobic oxidation of dichloroethene (Scherr et al., 2011) before, to contaminated aquifers. The addition or targeted stimulation of microorganisms equipped with the metabolic capability to transfer electrons to mineral surfaces, via secretion of chelators and microbial ES or expression of conductive protein appendages, such as *Geobacter* or *Shewanella*, represents another approach.

A tentative collection of bioremediation strategies incorporating extracellular electron transfer for petroleum hydrocarbon contaminated aquifer remediation is provided in Chapter 9.

3. Interactions in extracellular electron transfer

3.1. Possible participants and pathways

Different reactions can be mediated between participants of subsurface redox-processes via extracellular electron transfer, involving microbes as well as soluble and insoluble, organic and inorganic electron acceptors and donors. Extracellular electron transport is required in case the electron donor, electron acceptor or both can not be taken up by the bacterial cell, be it due to poor solubility, or if a direct contact between the microbe and electron sources or sinks is not possible, e.g. due to occlusion in pore spaces too small for microorganisms to enter. Possible extracellular pathways include, besides direct contact between cell and mineral surfaces:

- Secretion of chelators for, e.g. Fe(III) and Mn(IV) minerals
- Electron shuttling, i.e. reversible electron uptake and release by redox-active organic compounds excreted by microbes or in the natural background
- Expression of electrically conductive, pilus-like bacterial assemblages, also termed *nano-wires*

These will be discussed in more detail in Chapter 5.

3.2. Routes for extracellular electron transfer in the saturated zone

Electrons may be transported extracellularly between any of the parties participating in redox-processes: (i) microbes and poorly soluble bulk (ii) sinks and (iii) sources of electrons. Where either electron donor or acceptor are well soluble, usually no extracellular transport pathway is required. Figure 1 shows four possible flow paths or *routes* between the parties. Where extracellular transport is required, is indicated by solid lines. Dashed lines represent pathways for well-soluble participants.

Electron route A: electron donor to microbe

Via route **A** electrons are transported from an electron source to a microbe, a typical half-reaction pathway for the oxidation of zero-valent iron (ZVI). In case soluble polluting and non-polluting organic compounds (petroleum hydrocarbons, acetate and others) yield electrons, no extracellular pathway is required (dashed lines).

Route B: microbe to electron acceptor

In the oxidation of petroleum hydrocarbons coupled to iron or manganese reduction, the microbe gains energy from donating the electron obtained directly from the pollutant donor (dashed line to green microbe) to the energetically favourable, poorly soluble terminal electron acceptor (route **B**).

In case the terminal, inorganic electron acceptor is insoluble, as in the case for Fe(III) and Mn(IV) minerals, and/or no direct contact between cellular and electron accepting surface can be established, electron transfer rates can significantly be increased via extracellular electron transport vehicles (Jiang and Kappler, 2008). This can be aided by the production of electrically conductive pilus-like assemblages, also termed microbial *nanowires* (Reguera et al., 2005), electron shuttles or metal chelating agents chelators, as depicted in Figure 1.

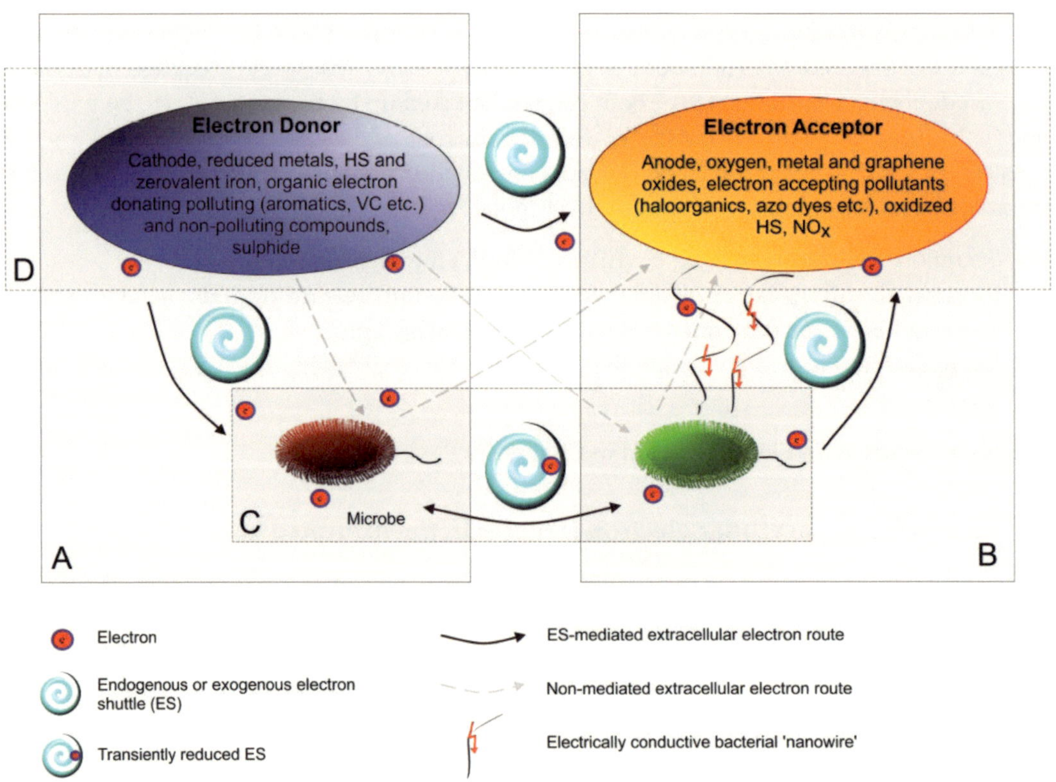

Figure 1. Electron routes in biogeochemical redox-processes: extracellular electron transport (A) from an electron source (donor) to microbe, (B) from microbe to electron sink (acceptor), (C) between microbes and (D) abiotic source/sink exchange. Solid lines indicate extracellular electron transport, dashed lines represent direct transport, usually via uptake of donor or acceptor. HS = humic substances, VC = vinyl chloride.

Route C: electron transport between microbes

Shuttling of electrons between microbes is represented by route **C**. Electron shuttles have been hypothesized to syntrophically link diverse organisms in nature (Lovley et al., 2004). This may occur, for example, between outer and inner layers of a biofilm (Lies et al., 2005). Intraspecies electron transfer has been observed with a naphthoquinone moiety as shuttle in the gastrointestinal tract (Yamazaki et al., 1999); its functionality in the subsurface has not been investigated yet.

Route D: electron transfer between abiotic parties

Moreover, electron shuttles may participate in purely abiotic electron transfer reactions between electron donors and acceptors (route **D** in Figure 1).

Route **D** symbolises the abiotic reduction of metals by microbially reduced humic substances (HS), in case the HS-oxidation process is literally *terminal*. In case of microbial 'regeneration', i.e. re-oxidation of HS, it serves as an electron shuttle in route **B**. Route **D** also describes abiotic chlorophenol degradation via ZVI, mediated by the presence of humic and fulvic acids (Kang and Choi, 2009), or shuttle-mediated abiotic reductive dechlorination mechanisms (Lee and Batchelor, 2002). This route also refers to the ability of natural organic matter to shuttle between oxidized and reduced forms of mercury that has only recently been recognized (Gu et al., 2011; Zheng et al., 2012).

4. Breathing solids: microbial respiration of poorly soluble iron and manganese minerals in uncontaminated environments

4.1. Not attached: respiration from a distance

Electrons need to be associated to a vector to participate in reactions beyond the immediate bacterial surface, i.e for travelling distances exceeding 0.01 μm (Gorby et al., 2006; Gray and Winkler, 1996). Soluble electron acceptors commonly used for petroleum hydrocarbon degradation include oxygen, nitrate and sulphate. These compounds can easily diffuse into the cell where they are reduced. Unlike these, the prevalent forms of iron and manganese in the environment are insoluble and remain extracellular, rendering the direct contact of cell and mineral surfaces an apparent necessity for respiration. This is contrasted by the ability of a number of microorganisms for dissimilatory reduction of undissolved mineral oxy(hydr)oxides, entailing electron transfer over distances of 50 μm and up. As an example the reduction of physically trapped or occluded iron(III) crystals was observed (Lies et al., 2005; Nevin and Lovley, 2002b). Different strategies enabling for long-distance electron transfer have been identified; they will be discussed in Chapter 5.

Microbes capable of dissimilatory metal reduction (DRMB) are phylogenetically dispersed throughout the *Bacteria* and *Archaea*; the most prominent members belong to *Geobacter, Shewanella* and *Geothrix* (see Chapter 7).

4.2. Biogeochemical role of poorly soluble iron minerals

Iron, comprising approximately 3.5% per mass of the earth's crust (Martin and Meybeck, 1979; Reimann and de Caritat, 2012) is the most abundant redox-active metal in the modern environment. While the main iron fraction is located in silica structures, approximately one third of iron atoms are present as amorphous or crystalline oxides in surface rocks (Canfield, 1997). At a circumneutral pH, iron is present primarily as quasi-insoluble minerals in di-valent ferrous (Fe(II), e.g. wüstite) trivalent ferric (Fe(III), e.g. goethite) or mixed-valence (Fe(II,II), e.g. magnetite) oxides, hydroxides and oxihydroxides (Scheffer and Schachtschabel, 2002).

Quasi-insoluble refers here to log K values of around -40 for goethite, ferric hydroxide and hematite (Morel and Hering, 1993). In the face of the poor solubility of iron oxides, it may come surprising that iron redox cycling in soil and sedimentary environments is now assumed to be dominated by microbial action (Arrieta and Grez, 1971; Chaudhuri et al., 2001; Weber et al., 2006). In fact, Fe(III) represents one of the most important terminal electron acceptors for microbes under anaerobic conditions (Lovley et al., 2004; Weber et al., 2006).

Iron, in its various oxidized or reduced states, has a high versatility in energy-creating processes in oxygen-rich and suboxic environments, as an electron donor for iron-oxidizing microbes under oxic and anoxic conditions as well as a terminal electron acceptor under anoxic conditions, and also influences other biogeochemical cycles. Microbial iron redox-cycling, also termed the *iron redox wheel*, is mechanistically connected to carbon, nitrogen, phosphorous and sulphur redox cycling (Li et al., 2012).

While poorly crystalline oxides appear to be readily metabolized (Phillips et al., 1993), the microbial utilization of highly crystalline oxides as electron acceptors in a (real) soil or sediment environment appears to be thermodynamically unfavourable (Thamdrup, 2000).

4.3. Dissimilatory manganese reduction

Mineral-bound manganese is significantly less abundant than iron in our environment, contributing approximately 0.07% by mass to surface rocks (Martin and Meybeck, 1979; Reimann and de Caritat, 2012).

The main fraction of manganese in the earth's surface, roughly 75%, is present as oxides (Canfield, 1997) with the remainder located in silicates. Minerals contain varying amounts of both Mn(III) and Mn(IV), with poorly crystalline manganite, vernadite and birnessite minerals as most abundant minerals (Friedl et al., 1997; Zhu et al., 2012). Similarly to iron cycling, biotic manganese oxidation and reduction are commonly encountered processes in soils and sediments. It appears that the mechanisms for manganese respiration and oxidation are similar to those for iron, as most organisms that can respire Fe can also respire Mn and vice versa (Lovley et al., 2004; Shi et al., 2007).

4.4. Behaviour of reduced manganese and iron

Following reduction, both Fe(II) and Mn(II) are re-oxidized biologically or chemically rather quickly. The thermodynamics of Fe(II) regeneration appear more favourable than those for

Mn(II) oxidation. Adding to this, Mn(II) has higher aqueous solubility than Fe(II), rendering microbe-driven post-depositional redistribution by reductive dissolution and transport from anoxic to oxic interfaces, where precipitation takes place, a main mechanism governing manganese localization (Thamdrup, 2000). Every atom may participate in subsequent reduction / oxidation cycles as many as 100 times (Thamdrup, 2000).

4.5. Side note: Bacterial reduction and oxidation of heavy metals

Besides iron and manganese, also polluting metals can be involved into bacterially mediated oxidation and reduction processes. These approaches can be used for bioremediation and the facilitated recovery of metals from waste water streams due to precipitation (Lovley and Phillips, 1992).

Uranium reduction

Some microbes can also reduce uranium (Cologgi et al., 2011; Lovley and Phillips, 1992; Lovley et al., 1991) at rates comparable to those of Fe(III) reduction (Lovley and Phillips, 1992). In contrast to manganese and iron, uranium precipitates during reduction from its hexavalent to its tetravalent form. The simultaneous precipitation of uranium with the reduction of Fe(III) has been observed for *Geobacter sulfurreducens* (Cologgi et al., 2011), leading to the assumption that similar mechanisms are governing both processes.

Reduction of other elements by DRMB

Similarly, *Shewanella* has been observed to precipitate reduced chrome (Cr(III)), likely involving the same enzymes as for Fe(III) reduction (Belchik et al., 2011). It can be hypothesised that the expression of extracellular electron transport pathways for soluble, oxidized electron acceptors precipitating during reduction, such as chrome and uranium was intended to keep the precipitates from forming inside the cells.

Dissimilatory iron-reducing bacteria were also found to participate in the reduction of technetium (Istok et al., 2004), cobalt, gold (Kashefi et al., 2001) and other metals, metalloids and radionuclides (Lloyd, 2003) as well as graphene oxide reduction (Jiao et al., 2011). These abilities may be used for the recovery or removal of metal resources from waste water streams or *in situ* remediation.

The ability to grow on these metals as sole electron acceptors has not been adequately demonstrated. Reducible iron and manganese minerals, however, will quantitatively dominate in most contaminated environments over heavy metals (Lovley et al., 2004) as terminal electron acceptors.

5. Microbial strategies to enhance extracellular electron transfer

In this chapter, mechanisms devised by DMRB for accessing poorly soluble electron acceptors are reviewed. Most of our understanding of these processes is derived from investigations of dissimilatory iron and manganese reduction under anaerobic conditions. *Geobacter* and *Shewanella* are considered as model organisms in this respect (cf. Chapter 7).

In the course of microbial energy generation, electrons are transferred from an electron source to a sink, usually inside a cell. DMRB, in contrast, are faced with the problem of how to effectively access an electron acceptor that can not diffuse to the cell. In case the electron donor or acceptor do not permeate through the membrane, as in the case of insoluble metals, or where the presence of redox-products inside the cell is not desired, such as for uranium precipitates, cells can rely on active processes to access poorly soluble TEAs or utilize extracellular electron shuttles. These are naturally present or anthropogenically administered compounds that have the potential to facilitate extracellular electron transfer.

Several mechanisms to overcome the physiological challenges for the dissimilatory reduction of poorly soluble electron acceptors have been identified.

5.1. Direct transport and electrochemical wiring

Direct electron transport

One route for extracellular electron transfer is the direct transport of electrons between the cellular envelope and electron accepting mineral surfaces, however requiring direct contact between cell and mineral surfaces. In these reactions, electrons have been found to travel over distances of 10 to 15 Å (Kerisit et al., 2007).

This direct transfer is mediated by outer-membrane cytochromes directly to the mineral's surface, and are typical of Gram-negative bacteria such as *Geobacter* and *Shewanella* (Hernandez and Newman, 2001). *Geobacter* and also *Shewanella* are representatives of a family of ferric iron reducers that were found in a variety of anaerobic environments (more about them in Chapter 7). The involved multihaem c-type cytochromes are essential electron-transferring proteins, rendering the journey of respiratory proteins from trough the cell to the outer membrane possible (Myers and Myers, 1992; Shi et al., 2007) and forming a respiratory chain extended beyond the cell's physiological periphery.

Directly nano-wired

Both *Shewanella* and *Geobacter* were observed to form pilus-like assemblages extending to the mineral surface when grown on insoluble electron acceptors, representing a special case of direct transport.

Geobacter was reported to produce monolateral pili to access iron and manganese oxides physically, i.e. via chemotaxis (Childers et al., 2002). Mineral reductases are located on the outer cell membrane but also on electrically conductive pilus-like assemblages prominently termed *nano-wires* (see Figure 1). These conductive protein nanostructures are 'wiring' the cell and the mineral phase, enabling for electron transport via a physical extension of the cell, thus outsourcing direct contact between cell and mineral surface to thin conductive appendages. Evidence for the production of *nano-wires* has been reported for both *Shewanella* (Gorby et al., 2006) and *Geobacter* (Reguera et al., 2005), however with distinct molecular mechanisms for the electron transport.

Different strategies are required for iron reducing microorganisms lacking conductive surface cytochromes, requiring for alternative strategies for extracellular electron transfer.

5.2. Electron shuttles and metal chelating agents

Electron shuttles are small molecules capable of undergoing repeated reduction and oxidation processes. An electron shuttle can be an organic or inorganic compound that is reversibly redox active and has the right redox potential, i.e. poised between the E_0' of the reductant and the oxidant. It serves as an electron acceptor and, once reduced, can itself transfer electrons to other organic or inorganic electron acceptors, whereupon it becomes re-oxidized. Such a shuttle provides a mechanism for an indirect reduction or oxidation process. In principle, a single shuttle molecule could cycle thousands of times and, thus, have a significant effect on the turnover of the terminal acceptor in a given environment.

Extracellular electron shuttles may be produced by microorganisms, originate from the humification of plant and animal matter or be anthropogenically added. While most electron shuttles are organic, heterocyclic aromatic structures, theoretically also inorganic, nonenzymatic compounds such as sulphide and reduced uranium, and organic compounds equipped with sulfhydryl groups (Nevin and Lovley, 2000) may serve as electron shuttles themselves.

In Figure 2 the redox-active groups of frequently encountered microbial, natural and anthropogenic electron shuttles are displayed. Heterocyclic aromatic compounds with redox-active functional groups, with di-ketone moieties in quinones, and nitrogen for phenazines and viologen, are common structures. Isoalloxazine represents the reactive structure in flavins.

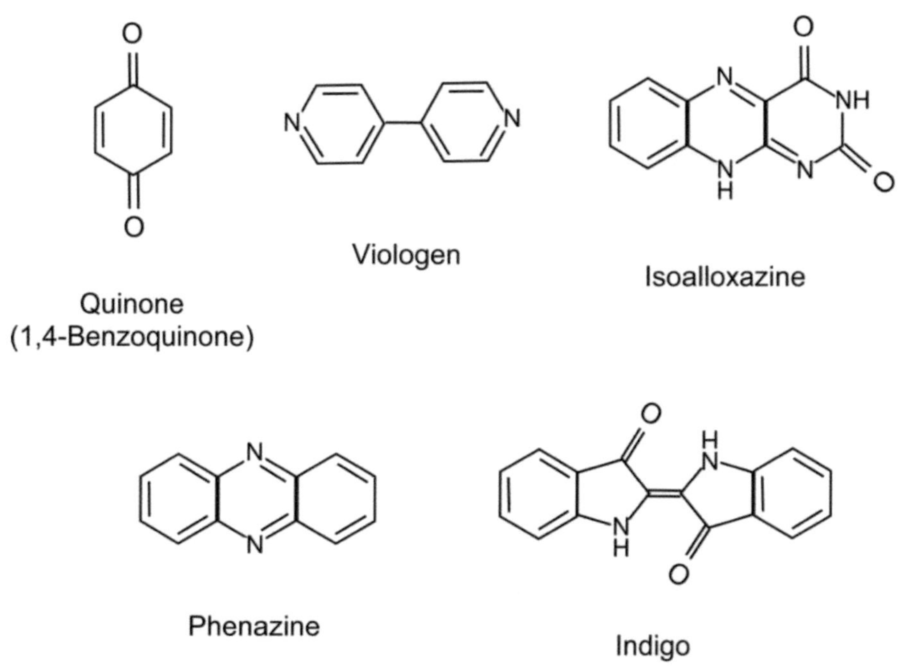

Figure 2. Main redox-active structures found in extracellular electron shuttles

5.2.1. Microbially produced extracellular electron transporters

In case the electron acceptor is physically detached from the cell, e.g. oxides occluded in pores, direct contact is not possible for a large fraction of bacteria, e.g. in biofilms, or the microbe lacks the enzymatic toolkit to respire electron acceptors, electrons are encouraged to take alternative routes through extracellular space by using transport facilities provided by the microorganism itself, such as electron shuttles and / or chelating agents.

The production of electron transport mediators is known for many microbes. Beside melanin, produced by *Shewanella alga* (Turick et al., 2002), phenazine derivatives (Hernandez et al., 2004), flavins (von Canstein et al., 2008), quinone-type structures (Nevin and Lovley, 2002a) and possibly also siderophores (Fennessey et al., 2010; Kouzuma et al., 2012) were shown to enhance anaerobic ferric iron respiration.

Phenazine and its derivatives

Phenazines are small redox-active molecules secreted by bacteria and functionally similar to those of anthraquinones. They are, due to their antibiotic effect, believed to be part of the organisms' biological warfare arsenal (Chin-A-Woeng et al., 2003; Hernandez and Newman, 2001). They are able to reductively dissolve iron and manganese (Hernandez et al., 2004; Wang and Newman, 2008), and are applied in microbial fuel cells (Pham et al., 2008). Phenazines are, however, of toxicological concern, such as pyocyanin, a blue pigment excreted by *Pseudomonas aeruginosa*, rendering the application of phenazines in environmental biotechnology an intricate task.

Flavins

Flavins have a midpoint potential of -0.2 to -0.25 V and can reduce most iron oxides and soluble forms of ferric iron, contributing to metal acquisition in plants and yeasts (Marsili et al., 2008). They have been proven to mediate the reduction of iron hydroxides inaccessible to microbes, via penetration through pores too small to allow for microbial passage (Nevin and Lovley, 2002b). Their isoalloxazine group acts as metal chelator.

The production of iron-solubilizing ligands has first been observed for *Shewanella putrefaciens* (Taillefert et al., 2007), which were shortly afterwards simultaneously identified as flavins by von Canstein and Marsili (Marsili et al., 2008; von Canstein et al., 2008). Also other γ-proteobacteria and *Marinobacter* are able to produce flavins, which may be used or *pirated* by other organisms directly (Hirano et al., 2003; von Canstein et al., 2008) and from flavin-coated surfaces (Marsili et al., 2008). The role of flavins in electron shuttling has recently been reviewed (Brutinel and Gralnick, 2012).

Side note: flavin or not flavin

Flavins were identified to have a major role in iron respiration of *Shewanella oneidensis*. In this context, both riboflavin (RF, Vitamin B_2) and flavin mononucleotide (FMN) were detected in iron-respiring *Shewanella* cultures. Flavins are now well-studied (Brutinel and Gralnick, 2012; Marsili et al., 2008; von Canstein et al., 2008) examples of endogenously synthesized electron transporting compounds deployed to donate respiratory electrons to a distant mineral phase. Here, electrons are conveyed to extracellular flavins via cytochromes that are re-oxidized upon direct contact to Fe(III), resulting in increased electron transfer rates than at direct contact to the mineral surface (Brutinel and Gralnick, 2012).

Experimental evidence for the functionality of flavin-mediated enhanced iron respiration was manifested in the presence of flavins in the supernatant of *Shewanella* reducing poorly soluble Fe(III). A stoichiometric ratio of 1 (FMN oxidized) to 2 (Fe(II) produced) was noted (von Canstein et al., 2008). The nature of the force bringing these compounds into the outer cell space remains disputed in that FMN and RF may either be actively secreted (Marsili et al., 2008; von Canstein et al., 2008) or merely be released during cell lysis as part of the common intracellular enzymatic mix. In this context, the former hypothesis is corroborated by different concentrations and species of intra- and extracellular flavins detected in the supernatant and the proportionality of extracellular flavin concentrations to live rather than dead cell density (von Canstein et al., 2008). These arguments were, in turn, challenged recently (Richter et al., 2012).

Siderophores

Siderophores are low molecular weight compounds produced by bacteria, fungi and plants when exposed to low-iron stress conditions, and are very strong Fe(III) chelating agents. Most of them contain catechol or hydroxamate groups as iron-chelating groups (Neilands, 1995). In forming Fe(III)-siderophore complexes, they are scavenging Fe(III) from the environment, rendering it more accessible for bacterial uptake. Its role in increasing metal respiration is, to our knowledge, uncertain.

Although siderophores were found to catalyze the mineralisation of carbon tetrachloride, it is not known whether microorganisms can gain energy from the process (Lee et al., 1999). Recent research suggests indirect effects of siderophores on respiration in increasing iron uptake, which is in turn required for the synthesis of cytochromes for manganese respiration (Kouzuma et al., 2012), while others indicate no effect of siderophore-related iron solubilisation on its respiration (Fennessey et al., 2010). It appears likely that iron-containing shuttles participate in iron solubilizing mechanisms, where the cation remains with the redox-active centre, thus corresponding to permanent rather than transient iron complexation.

5.3. Exogenous electron shuttles and chelators

5.3.1. *Types of electron shuttles*

For a large variety of naturally occurring redox-active compounds, at least some functionality as extracellular electron shuttles was described (Guo et al., 2012; O'Loughlin, 2008; Van der Zee and Cervantes, 2009; Watanabe et al., 2009; Wolf et al., 2009). Main functional structures are depicted in Figure 2. Frequently encountered shuttles encompass the following:

- Quinones, including
  - anthraquinones, their sulfonates, carboxylic acids and chlorinated moieties

  - juglone,
  - lawsone,
  - menadione and other naphthoquinones
  - ubiquinone, coenzyme Q_{10}
- Humic substances
- Phenazines cores, including
  - neutral red
  - methylene blue
- Vitamine cores, including
  - corrin
  - isoalloxazine
- Indigo sulfonates and other derivatives
- Viologens, including
  - methyl viologen
  - benzyl viologen
- Cysteine and other thiol-containing molecules

Amongst these, the role of quinones and humic substances are discussed in more detail below.

5.3.2. Quinones

Structure

Chemically, quinones are fully conjugated cyclic di-ketone structures derived from the oxidation of aromatic compounds, where an even number of –CH= groups is converted into –C(=O) groups. This is adjoined by a re-arrangement of double bonds and connected to the loss of aromaticity. Quinones may also be considered as products from polyphenol reduction (Figure 3). Quinones are stable throughout a number of repeated oxidation and reduction processes. In their redox-cycling, one- and two-electron reduction reactions lead to the formation of free semiquinone radicals and hydroquinones, as displayed in Figure 3 (Michaelis and Schubert, 1938).

Occurrence

Quinones are abundant structures in nature, participating in a wide variety of reactions in pro- and eukaryotic organisms. Membrane-bound ubiquinones and menaquinones shuttle electrons between respiratory protein complexes in bacteria. Phylloquinone participates in the electron transfer chain of in photosynthesis. Other quinones may have antimicrobial (Lown, 1983; Zhou et al., 2012) or even toxic (Shang et al., 2012) effects, such as juglone. Isoalloxazine, the reactive structure in flavins, is also a quinone (Muller et al., 1981).

Role of quinones in electron shuttling

The hydroquinone moiety chemically reduces Fe(III) to Fe(II) while it is simultaneously oxidized to a quinone (Scott et al., 1998). In turn, it is reduced back to hydroquinone by microbes deriving electrons from hydrocarbon oxidation and other processes. This is a type

B reaction (Figure 1), with the electron donor being a hydrocarbon and the acceptor an insoluble mineral, such as a Fe(III) or Mn(IV) mineral. Although the energy gain from quinone reduction is lower than from direct reduction of Fe(III) minerals, the use of extracellular shuttles enables for significantly increased reduction rates, i.e. a higher bacterial energy gain per unit time (Jiang and Kappler, 2008).

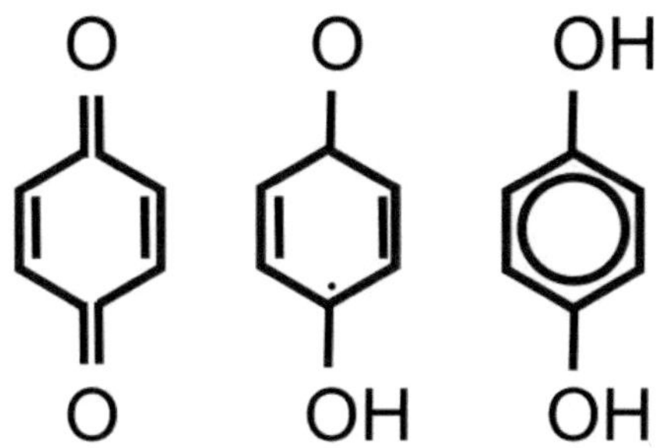

Figure 3. Simple quinoid structures: 1,4- or para-benzoquinone (left), its semiquinone (center) and its hydroquinone (right)

The manifestation of a type **A** route would reflect electron shuttling from an organic electron donor to a halogenated or nitrous contaminant.

In summary, electron shuttles participate in facilitating the electron transfer from the microbe towards the electron acceptor in the type **B** reaction and from electron donor to microbe for sulphide, reduced metals or the cathode for a type **A** reaction; in both contexts, quinones are often encountered.

Iron-reducing microorganisms are able to grow on extracellular quinones as the sole source of energy. Humic substance-derived quinones are possible the most important natural source of extracellular shuttles for Fe(III) reducers (Lovley et al., 1996a; Scott et al., 1998).

In many studies, anthraquinone disulfonate (AQDS) was used as a model humic substance (Aulenta et al., 2010; Bhushan et al., 2006; Collins and Picardal, 1999; Kwon and Finneran, 2006; Kwon and Finneran, 2008; Scherr et al., 2011) to study extracellular electron transfer in the reduction of halogenated and nitrous organic compounds.

5.3.3. Humic substances

Humic substances (HS) are a heterogeneous family of dark-coloured, biologically largely refractory compounds originating from the incomplete biodegradation of plant, animal and microbial debris. They account for up to 80% of soil and sedimentary organic carbon (Schnitzer, 1989). Humic substances consist of structurally diverse macromolecules, containing a variety of functional groups including hydroxy- and carboxygroups as well as aromatic and aliphatic structures that can be resolved using infrared and magnetic resonance spectroscopy, amongst others (Ehlers and Loibner, 2006; Nguyen et al., 1991).

On a more general scale, the classification of humic substances is commonly performed based on isolation procedures rather than on molecular characteristics, usually by their acid (in)solubility, with brown to black *humic acids* (HA) defined as insoluble at a pH <2, with yellow-coloured *fulvic acids* (FA) completely soluble and *humins* insoluble both under all pH conditions (Scheffer and Schachtschabel, 2002).

Multiple interactions occur between the organic and inorganic soil matter fractions, including encapsulation (Baldock and Skjemstad, 2000), formation of adducts with organic contaminants (Karickhoff et al., 1979), chelation of metals (Arrieta and Grez, 1971; Gu et al., 2011; Lovley et al., 1996b), and as terminal and subterminal electron acceptors for abiotic and biotic reduction processes (Benz et al., 1998; Jiang and Kappler, 2008; McCarthy and Jimenez, 1985), amongst others. The multiple functions of humic substances have been proposed to be incorporated in different remediation technologies for contaminated sites, including bioremediation, reactive barriers and *in situ* immobilization (Perminova and Hatfield, 2005).

In terms of contributing to extracellular electron transfer enhancing contaminant transfer, humic substances play multiple roles.

Humic acids as chelators

Humic acids were found to increase the concentration of soluble Fe(III) by forming chelates (Lovley et al., 1996b). Chelated Fe(III) has a higher redox potential, rendering Fe(III) reduction more thermodynamically favourable (Thamdrup, 2000) and rendering the cation more accessible to microbes for reduction (Dobbin et al., 1995). It is, however, not clear whether extracellular chelated Fe(III) can be used by the cell directly via surface contact or absorption by the periplasm is required prior to reduction (Haas and Dichristina, 2002; Lovley et al., 1996b).

Terminal electron acceptors versus electron shuttles

Although humic substances are rather refractory to microbial breakdown, microorganisms interact with a variety of their functional groups, including the donation of electrons to reactive humic moieties. The ability to reduce humic substances and iron by bacteria and archaea appears to be closely linked (Benz et al., 1998; Lovley et al., 1998). The nature of electron transfer to humic or fulvic acids may be terminal or subterminal. In the latter case, microbially reduced humic matter abiotically transfers electrons to the Fe(III) surface, being re-oxidized themselves in the process (Lovley et al., 1996a; Lovley et al., 1996b) and thus function as natural, biogenic extracellular electron shuttles (see Figure 1). Reduced humic matter can also be „pirated" by other microorganisms (Lovley et al., 1999), i.e. those that did not create them (route **C** in Figure 1) and be used as electron donors for microbially mediated reactions (route **A**).

Originally, the effect of humic substances in promoting biodegradation of organic pollutants was assumed to be dominated by chelation (Lovley et al., 1996b). Humic substances, however, are only weak chelators of iron. The role of humified organic matter as electron shuttle was first raised in 1989 (Tratnyek and Macalady, 1989) and is also referred to as the *Tratnyek-Macalady hypothesis* (Sposito, 2011). In fact, the function of fulvic acids of different origin can be explained as an initial contribution by electron shuttling, further enhanced by iron complexation in latter stages of contact (Royer et al., 2002).

Quality and origin of humic matter

Different humic substance fractions and origins have different effects on the rate and extent of extracellular electron shuttling. While there might be no effect at all by neither humic nor fulvic acids (O'Loughlin, 2008), humic acids, on a mass unit base, were found to stimulate iron reduction more strongly stronger than fulvic acids, also in environmentally relevant concentrations, due to thermodynamic mechanisms (Wolf et al., 2009). On the other hand, aquatic humic substances appear to have a higher electron accepting capacity than those derived from terrestrial environments (Royer et al., 2002; Scott et al., 1998). On a general basis, the shuttling capacity of humic matter is determined mainly by with its aromaticity (Aeschbacher et al., 2010; Chen et al., 2003), where carbohydrate fractions were found to be less redox-active than polyphenolic fractions (Chen et al., 2003)

Soil organic matter aromaticity is also one of the main functional groups in determining sorption of petroleum hydrocarbons (Ehlers et al., 2010; Perminova et al., 1999). Surface-adsorbed and freely dissolved humic matter were found to exhibit similar shuttling activities (Wolf et al., 2009).

The disputed role of quinones in electron shuttling by humic substances

While the role of humic substances in soil and sediment electron shuttling is unchallenged, the nature of functional structures in humic or fulvic acids conveying the electron shuttling capacity to humic substances is not fully elucidated yet.

Profound influence of humic quinone moieties in conveying shuttling properties to humic substances has been attested by numerous studies (Aeschbacher et al., 2010; Jiang and Kappler, 2008; Kang and Choi, 2009; Perminova et al., 2005; Scott et al., 1998). Quinone moieties such as AQDS are often used as practical and standardized surrogates to model the effect of humic substances in extracellular electron transfer, while representing only the humics' redox properties, since AQDS can not chelate metals. The relevance of quinones, however, can be demonstrated by comparing shuttling efficacy of humic acid and pure AQDS. Two gram per litre of humic acids had a similar stimulating effect on the reduction of poorly crystalline ferric iron as approximately 40 mg / L AQDS (Lovley et al., 1996a).

In the search for other redox-active groups in humic substances, one model proposed three distinct sites participating in the redox activity of humic and fulvic acids (Ratasuk and Nanny, 2007), including two quinone structures with different electronegative properties plus one non-quinone structure. Evidence was supplied in that humic substances artificially depleted of their aromatic structures were not entirely depleted of their reducing capacity. In a different study, the humic substances' content of free radicals accounted for only a small fraction of the observed electron equivalents in shuttling between iron and iodine moieties (Struyk and Sposito, 2001).

These studies corroborate the possibility that also non-quinone structures participate in conveying redox-activity to natural organic matter. The nature of these compounds, however, is unknown, while there is general agreement points towards that quinone moieties are the *main*, however not only redox-active groups in humic substances.

Further understanding can be expected by the computational and procedural determination of quinone redox properties (Aeschbacher et al., 2010; Cape et al., 2006; Guo et al., 2012).

5.4. Concentrations of electron shuttles

Depending on the stability of the electron shuttling compound, it may undergo multiple cycles of electron take up and release. Thus, concentrations of redox-active moieties in the environment theoretically need to be substoichiometric in comparison to the amount of electron donating hydrocarbon and acceptor.

The oxidation state of a mediator itself also, at least theoretically, plays a role in its efficacy. Thus, an electron shuttle is effective at certain ratios of oxidized : reduced between 1:100 and 100:1 (Meckstroth et al., 1981), although in many studies it is attempted to completely reduce shuttles using harsh methods ahead of experimental use.

Practically, concentrations of pure quinones generally between 0.1 to 100 micromoles are commonly used for shuttling experiments. For some compounds, however, higher concentrations are inhibitory due to antimicrobial effects (Lown, 1983; Wolf et al., 2009). For some of the tested substances tested in a recent study (Wolf et al., 2009), a linear correlation between the normalized reaction rates and the logarithm of their concentrations was observed (humic and fulvic acid and AQDS), while others showed no concentration dependence (carminic acid and alizarin).

There has been criticism that concentrations of humic and fulvic acids commonly employed for experimental shuttling studies would far exceed those relevant under natural conditions, which are in the milligram per litre scale (Wolf et al., 2009), amounting to between 5 and 25 mg C/L (Curtis and Reinhard, 1994; Jiang and Kappler, 2008; O'Loughlin, 2008).

For humic and fulvic acids, enhancing effects on Fe(III) reduction attributed to electron shuttling were observed for concentrations of dissolved fulvic acids as low as 0.61 mg/L and of humic acids of 0.025 mg/L, i.e. at concentrations which do occur under natural conditions (Wolf et al., 2009) such as in pore waters of sediments rich in organic matter.

6. Thermodynamic considerations for extracellular electron shuttles

The ability to shuttle electrons between different organic and inorganic participants in the subsurface is largely dependent on the energies driving the extracellular electron transfer processes in simultaneous oxidation and reduction processes.

6.1. Minimum requirements for ATP production

The relation of electrochemical potentials of electron donor, shuttle and electron acceptor determine the amount of energy, in terms of ATP generation or free energy ($\Delta G^{0\prime}$), a microorganism may harvest from the electron transfer. A minimum energy yield of -20kJ/mol is required for ATP production (Schink, 1997). This determines the lower threshold amount of energy required to be transformed in the process of microbial electron transfer from donor to shuttle to allow for ATP synthesis in the cells (Wolf et al., 2009). This first step depends on the potential of the electron donor or, more precisely, of its redox couple, and the shuttle.

In the subsequent, rate-limiting step of electron transport from the shuttle to the electron acceptor, the difference voltage between the mediator and the acceptor couple should be as negative as possible (Wolf et al., 2009). From a theoretical viewpoint, a mediator, assuming a concentration of 1M/L, can efficiently mediate between compounds with a $E^{0'}$ range of +/- 118 mV of its own midpoint potential (Meckstroth et al., 1981).

These relations should be regarded in the selection of an electron shuttle. Several publications are dedicated to the description of the redox-properties of a wide range of mediators (Bird and Kuhn, 1981; Fultz and Durst, 1982; Meckstroth et al., 1981).

6.2. Effect of shuttle structure on its midpoint potential

The location and type of the mediator's functional groups influences its potential. As an example, quinones equipped with an increasing number electron donating groups (e.g. –CH_3) become less easily reduced. Redox-potentials (all $E^{0'}$) decrease from 280 mV for the unalkylated *para*-benzoquinone to 180 mV for 2,5-*p*-benzoquinone and to 5mV for the fully alkylated congener (2,3,5,6-tetramethyl-*p*-benzoquione). On the other hand, electron-withdrawing sulphone groups facilitate reduction, as is reflected by an increasing potential along with increasing sulphonation for anthraquinone (AQ). Here, an increase of the compounds' potential from -225 mV for AQ-2-sulphonate by almost 40 mV to -184 mV for AQ-2,6-disulphonate is connected to the addition of one sulphone group (Fultz and Durst, 1982). The location of electron donating and accepting groups in respect to the keto-groups also determines the shuttles' stability in recurring redox-processes (Watanabe et al., 2009).

6.3. Practical relevance

In practice, however, the translation of thermodynamic principles to complex biogeochemical interactions can seldom be expected to be straightforward. In this case, small-scale inductive effects do not manifest themselves in decisively different shuttling efficacies, however a good correlation of shuttle electrochemical potential to Fe(II) production can be observed (O'Loughlin, 2008; Wolf et al., 2009). Mediation of ferrihydrite reduction with formate as electron donor was found to be most effective for a quinone range with potentials (all $E^{0'}$) of between -137 mV (2-hydroxynaphtoquinone) to -225 mV for AQS as mentioned above (Wolf et al., 2009). These potentials are closely related to those of flavins excreted by *Shewanella*, around -215 mV (Marsili et al., 2008; von Canstein et al., 2008). Thermodynamic issues of extracellular electron transfer are discussed elsewhere in more detail (O'Loughlin, 2008; Watanabe et al., 2009; Wolf et al., 2009).

7. *Geobacter* and *Shewanella*: Well-studied iron and manganese breathers

Dissimilatory iron (III) reducing microorganisms are phylogenetically diverse and can be found within the bacteria and archaeal domains. They are able to gain energy from the reduction of Fe(III) on the cost of the oxidation of organic substances or molecular hydrogen. Behind their rather modest designation as iron-reducers the ability to participate in a great variety of biogeochemical processes is hiding. Mechanisms required for iron reduction also negotiate interactions with other organic and inorganic chemical species in the subsurface. Thus, the influence of these organisms and their metabolic products extends far beyond biogeochemical carbon and iron cycling, but also in dissolution of inorganic polluting and non polluting metals, transition metals, micro- and macronutrients and of organic pollutants. Phylogenetically distinct DMRB have different mechanisms to access insoluble electron acceptors. It appears that the ability to reduce these acceptors has evolved in evolution several times (Lovley et al., 2004).

Amongst the variety of involved microorganisms in the terrestrial and aquatic environment, the selection of microbial study objects is somewhat biased by their cultivability as pure cultures under laboratory conditions (Amann et al., 1995). The large diversity of DMRB shall not be underestimated by the selection of few cultivable organisms, which represent less than one percent of total bacterial cell counts in soil and sediments (Jones, 1977; Torsvik et al., 1990).

Beside numerous other organisms capable of dissimilatory metal and transition-metal reduction, including *Geothrix* (Coates et al., 1999) and reservoir-borne bacteria (Greene et al., 1997; Greene et al., 2009), the most well studied DMRB belong to *Geobacter* and *Shewanella* species.

7.1. Members of the *Geobacter* genus

Geobacter are a genus of δ-proteobacteria. They are gram-negative and chemoautotrophic strict anaerobes and were found to dominate iron reducing populations in hydrocarbon-polluted environments over other known Fe(III) reducers such as *Shewanella, Geothrix* and *Variovorax*. The latter appear to be unable to couple Fe(III) reduction to pollutant oxidation (Rooney-Varga et al., 1999; Snoeyenbos-West et al., 2000).

Thus, *Geobacter* species appear to be the primary agents in Fe(III) and Mn(IV) reduction coupled to the oxidation of organic compounds in anoxic terrestrial environments.

The first description of the Geobacter species was published by D. Lovley in 1987 (Lovley et al., 1987) after isolating them from a Potomac river sediment. The species G. *metallireducens* and *sulfurreducens* were described in 1993 (Lovley et al., 1993) and 1994 (Caccavo et al., 1994), respectively. Geobacter *sulfurreducens* can not, in contrast to G. *metallireducens*, reduce Mn(IV), and can not use alcohols or aromatic compounds as electron donors.

Geobacter can use AQDS as extracellular electron shuttle to increase electron transfer to poorly soluble electron acceptors, and can also precipitate Uranium (VI) (Cologgi et al., 2011). G. *metallireducens* is chemotactic towards the soluble reduction products (Fe(II) and Mn(II)) of iron and manganese minerals (Childers et al., 2002)).

Nano-wiring and charging

Geobacter form pilus-like appendages, so-called bacterial nanowires (Reguera et al., 2005), when grown on insoluble Fe(III) and Mn(IV) oxides, to transfer electron to distant and insoluble electron acceptors and electrodes. These pili are electrically conductive. This is independent from c-type cytochromes, which are, in contrast, responsible for terminal connections with the electron accepting surfaces. They are also responsible for microbial 'capacitation', which allows *Geobacter* to extracellularly store electrical charge in times of scarceness of electron acceptors (Lovley et al., 2011).

7.2. Members of the *Shewanella* genus

Members of *Shewanella*, originally isolated from dairy products and introduced as *Achromobacter putrefaciens* by Derby & Hammer in 1931, are gram-negative and belong to the class of γ-proteobacteria (Venkateswaran et al., 1999).

They are facultative anaerobic bacteria equipped with a high respiratory versatility, amongst others capable of using oxygen, nitrate, volatile fatty acids and Fe(III), Mn(IV), As(V) and U(VI) as electron sinks (Bencheikh-Latmani et al., 2005; Cruz-García et al., 2007; Lim et al., 2008; Lovley et al., 2004; Myers and Nealson, 1988; Venkateswaran et al., 1999; von Canstein et al., 2008), and supplemented by evidence for Cr(VI) respiration (Bencheikh-Latmani et al., 2005). Extracellular respiration by *Shewanella oneidensis* has also been noted to comprise graphene oxide (Jiao et al., 2011).

Flavins, electron shuttles and nano-wires

The versatility in using electroactive surfaces to donate electrons is reflected in the use of anodes (Logan, 2009), and also the ability to produce electrically conductive nanowires, similar to *Geobacter*, themselves (Gorby et al., 2006).

Shewanella are known to excrete quinones (Newman and Kolter, 2000) that carry electrons from the cell surface to the electron acceptor that is located at a distance from the cell. Excreted flavins chelate Fe(III), rendering it more accessible to the cells (von Canstein et al., 2008), thus promoting Fe(III) reduction at a distance.

8. Anaerobic hydrocarbon oxidation coupled to Fe(III) and Mn(IV) reduction

8.1. Naturally mediated extracellular electron transfer

A variety of easily degradable electron donors can be used for the anaerobic reduction of Fe(III) and Mn(IV), including short chain organic acids, alcohols or sugars (Lovley et al., 1996a). Iron- and manganese reducers are also known to dwell in petroleum reservoirs (Greene et al., 1997; Greene et al., 2009). This suggests that the metabolic pathways for the reduction of poorly soluble electron acceptors are resistant to exposure to potential inhibitors of microbial activity; this is of potential concern since low molecular weight hydrocarbons are known to act as disruptors of biological membranes.

In fact, poorly soluble electron acceptors can also be used for the anaerobic oxidation of less readily degradable organic compounds, including petroleum hydrocarbons. Several studies documented the occurrence of microbial hydrocarbon degradation coupled to the reduction of Fe(III) and Mn(IV) as sole electron acceptors with aquifer and sedimentary materials. Biodegradation of petroleum hydrocarbons under iron- and manganese reducing conditions has been noted primarily for low molecular weight aromatic compounds. This includes benzene (Kunapuli et al., 2008), toluene (Coates et al., 1999; Edwards and Grbic-Galic, 1994; Langenhoff et al., 1997; Lee et al., 2012), ethylbenzene (Siegert et al., 2011) and *o*-xylene (Edwards and Grbic-Galic, 1994) as well as phenol (Lovley et al., 1989), naphthalene and also liquid alkanes (Siegert et al., 2011).

It can be assumed that natural mechanisms, including chemotaxis, expression of electrically conductive pilus-like assemblages, extracellular chelators and molecular electron shuttles contribute to the occurrence of these processes.

8.2. Role of anthropogenic extracellular electron shuttles

The enhancement of naturally occurring iron and manganese reduction coupled to hydrocarbon oxidation is a promising strategy for petroleum hydrocarbon bioremediation (Lovley, 2011). There is, to our knowledge, only a limited amount of studies where such a strategy was followed. Humic substances and quinones were successfully applied to stimulate the degradation of a variety of aromatic compounds by Cervantes and co-authors, including the increase of toluene oxidation coupled to manganese respiration (Cervantes et al., 2001). Degradation of the fuel additive methyl tert-butyl ether (MTBE) under iron-reducing conditions was stimulated using humic substances (Finneran and Lovley, 2001) as well as quinones for anaerobic toluene oxidation (Evans, 2000). Iron chelation with humic substances increased anaerobic benzene oxidation (Lovley et al., 1996b), however a contribution by electron shutting was not discerned.

In the light of the large variety of potential applications of these processes for petroleum hydrocarbon bioremediation, more research tackling upon a broader variety of priority pollutants will be performed.

9. Outlook: Potential exploitation of enhanced extracellular electron transfer for *in situ* petroleum hydrocarbon bioremediation

The application of electron shuttles as amendments in a variety of biotechnological processes has gained increased attention in the past 20 years. Extracellular electron shuttles have, for example, been shown to enhance the *reductive* degradation of a large variety of organic contaminants, including azo- and nitro- compounds such as dyes and explosives, of chlorinated hydrocarbons, and to induce the precipitation of metals such as uranium and chromium (reviewed by Field et al., 2000; Hernandez and Newman, 2001; Van der Zee and Cervantes, 2009; Watanabe et al., 2009). Recent developments include the application of novel electron shuttles and the continued investigation to predicting the redox-properties the most complex natural redox mediators, humic substances (Aeschbacher et al., 2010; Kumagai et al., 2012). The exploitation of extracellular electron transfer in anaerobic petroleum hydrocarbon bioremediation is, in contrast, explored to a lesser extent.

The limited availability of electron acceptors constraining oxidative hydrocarbon biodegradation *in situ* can be circumvented by a variety of measures, including *biosparging*, the introduction of oxygen releasing compounds or the use of naturally occurring, well soluble alternative electron acceptors such as nitrate or sulphate.

Moreover, the abundance of solid-phase, however microbially accessible electron acceptors in the subsurface offers the potential for a significant decrease in the amount of reagents to be introduced into an aquifer, provided solid-phase electron acceptors can be efficiently involved in oxidative contaminant biodegradation.

The enhancement of microbially mediated mechanisms to enhance the accessibility of Fe(III) and Mn(IV) can be applied in a variety of engineered approaches for the biodegradation of aquifers contaminated with fuel hydrocarbons, tar oils and other contaminants susceptible to oxidative biodegradation.

The list of possible approaches includes the following:

i. Injection of electron shuttles carefully designed by type and concentration, into contaminated aquifers, to increase the available pool of native or resident electron acceptors such as Fe(III) and Mn(IV) for microbial hydrocarbon oxidation.
ii. Addition of electron shuttles to increase the electron accepting function of pre-existing engineered subsurface structures, such as permeable reactive barriers, gates and reactive zones.
iii. Use of natural or engineered materials slowly releasing electron shuttles by leaching, such as humic substances.
iv. Engineered approaches to increase the density and activity of microbial communities with the intrinsic capability to reduce poorly soluble native sources of electron acceptors, including but not limited to *Geobacter* species.

These considerations shall mediate research and development of an ever broader variety of potential opportunities to enhance naturally occurring oxidative contaminant degradation under suboxic conditions. The range of available tools for dealing with hydrocarbon contaminated environments may be extended by an additional instrument, the utilization of extracellular electron transport mechanisms designed by nature.

Abbreviations

ATP	adenosine-5'-triphosphate
AQDS	anthraquinone disulfonate
Cr	chrome
DMRB	dissimilatory metal reducing bacteria
ES	electron shuttle
FA	fulvic acids
Fe	iron
FMN	flavin mononucleotide

HA	humic acids
HS	humic substances
Mn	manganese
RF	riboflavin
TEA	terminal electron acceptor
ZVI	zero-valent iron

Author details

Kerstin E. Scherr
University of Natural Resources and Life Sciences Vienna,
Remediation Biogeochemistry Group, Institute for Environmental Biotechnology, Tulln
Austria

Acknowledgement

Financial support for this publication was provided by the Austrian Federal Ministry of Agriculture, Forestry, Environment and Water Management and the Government of Upper Austria, Directorate for Environment and Water Management, Division for Environmental Protection.

10. References

Aeschbacher, M., Sander, M., Schwarzenbach, R.P., 2010. Novel electrochemical approach to assess the redox properties of humic substances. Environmental Science and Technology 44, 87-93.

Aichberger, H., Nahold, M., Mackay, W., Todorovic, D., Braun, R., Loibner, A.P., 2007. Assessing natural biodegradation potential at a former oil refinery site in Austria. Land Contamination and Reclamation 15, 1-14.

Amann, R.I., Ludwig, W., Schleifer, K.H., 1995. Phylogenetic identification and in situ detection of individual microbial cells without cultivation. Microbiological Reviews 59, 143-169.

Arrieta, L., Grez, R., 1971. Solubilization of Iron-Containing Minerals by Soil Microorganisms. Applied Microbiology 22, 487-490.

Aulenta, F., Maio, V.D., Ferri, T., Majone, M., 2010. The humic acid analogue antraquinone-2,6-disulfonate (AQDS) serves as an electron shuttle in the electricity-driven microbial dechlorination of trichloroethene to cis-dichloroethene. Bioresource Technology 101, 9728-9733.

Baldock, J.A., Skjemstad, J.O., 2000. Role of the soil matrix and minerals in protecting natural organic materials against biological attack. Organic Geochemistry 31, 697-710.

Bartha, R., 1986. Biotechnology of petroleum pollutant biodegradation. Microbial Ecology 12, 155-172.

Belchik, S.M., Kennedy, D.W., Dohnalkova, A.C., Wang, Y., Sevinc, P.C., Wu, H., Lin, Y., Lu, H.P., Fredrickson, J.K., Shi, L., 2011. Extracellular Reduction of Hexavalent Chromium by Cytochromes MtrC and OmcA of Shewanella oneidensis MR-1. Applied and Environmental Microbiology 77, 4035-4041.

Bencheikh-Latmani, R., Williams, S.M., Haucke, L., Criddle, C.S., Wu, L., Zhou, J., Tebo, B.M., 2005. Global transcriptional profiling of Shewanella oneidensis MR-1 during Cr(VI) and U(VI) reduction. Applied and Environmental Microbiology 71, 7453-7460.

Benz, M., Schink, B., Brune, A., 1998. Humic acid reduction by Propionibacterium freudenreichii and other fermenting bacteria. Applied and Environmental Microbiology 64, 4507-4512.

Bhushan, B., Halasz, A., Hawari, J., 2006. Effect of iron(III), humic acids and anthraquinone-2,6-disulfonate on biodegradation of cyclic nitramines by Clostridium sp. EDB2. Journal of Applied Microbiology 100, 555-563.

Bird, C.L., Kuhn, A.T., 1981. Electrochemistry of the viologens. Chemical Society Reviews 10, 49-82.

Bregnard, T.P.A., Höhener, P., Zeyer, J., 1998. Bioavailability and biodegradation of weathered diesel fuel in aquifer material under denitrifying conditions. Environmental Toxicology and Chemistry 17, 1222-1229.

Brutinel, E.D., Gralnick, J.A., 2012. Shuttling happens: Soluble flavin mediators of extracellular electron transfer in Shewanella. Applied Microbiology and Biotechnology 93, 41-48.

Caccavo, F., Lonergan, D.J., Lovley, D.R., Davis, M., Stolz, J.F., McInerney, M.J., 1994. Geobacter sulfurreducens sp. nov., a hydrogen- and acetate-oxidizing dissimilatory metal-reducing microorganism. Applied and Environmental Microbiology 60, 3752-3759.

Canfield, D.E., 1997. The geochemistry of river particulates from the continental USA: Major elements. Geochimica et Cosmochimica Acta 61, 3349-3365.

Cape, J.L., Bowman, M.K., Kramer, D.M., 2006. Computation of the redox and protonation properties of quinones: Towards the prediction of redox cycling natural products. Phytochemistry 67, 1781-1788.

Cermak, J.H., Stephenson, G.L., Birkholz, D., Wang, Z., Dixon, D.G., 2010. Toxicity of petroleum hydrocarbon distillates to soil organisms. Environmental Toxicology and Chemistry 29, 2685-2694.

Cerniglia, C.E., 1992. Biodegradation of polycyclic aromatic hydrocarbons. Biodegradation 3, 351-368.

Cervantes, F.J., Dijksma, W., Duong-Dac, T., Ivanova, A., Lettinga, G., Field, J.A., 2001. Anaerobic Mineralization of Toluene by Enriched Sediments with Quinones and Humus as Terminal Electron Acceptors. Applied and Environmental Microbiology 67, 4471-4478.

Chaudhuri, S.K., Lack, J.G., Coates, J.D., 2001. Biogenic Magnetite Formation through Anaerobic Biooxidation of Fe(II). Applied and Environmental Microbiology 67, 2844-2848.

Chen, J., Gu, B., Royer, R.A., Burgos, W.D., 2003. The roles of natural organic matter in chemical and microbial reduction of ferric iron. Science of The Total Environment 307, 167-178.

Childers, S.E., Ciufo, S., Lovley, D.R., 2002. Geobacter metallireducens accesses insoluble Fe(iii) oxide by chemotaxis. Nature 416, 767-769.

Chin-A-Woeng, T.F.C., Bloemberg, G.V., Lugtenberg, B.J.J., 2003. Phenazines and their role in biocontrol by Pseudomonas bacteria. New Phytologist 157, 503-523.

Coates, J.D., Ellis, D.J., Gaw, C.V., Lovley, D.R., 1999. *Geothrix fermentans gen. nov., sp. nov.*, a novel Fe(III)-reducing bacterium from a hydrocarbon-contaminated aquifer. International Journal of Systematic and Evolutionary Microbiology 49, 1615-1622.

Collins, R., Picardal, F., 1999. Enhanced anaerobic transformations of carbon tetrachloride by soil organic matter. Environmental Toxicology and Chemistry 18, 2703-2710.

Cologgi, D.L., Lampa-Pastirk, S., Speers, A.M., Kelly, S.D., Reguera, G., 2011. Extracellular reduction of uranium via Geobacter conductive pili as a protective cellular mechanism. Proceedings of the National Academy of Sciences 108, 15248-15252.

Cruz-García, C., Murray, A.E., Klappenbach, J.A., Stewart, V., Tiedje, J.M., 2007. Respiratory nitrate ammonification by Shewanella oneidensis MR-1. Journal of Bacteriology 189, 656-662.

Curtis, G.P., Reinhard, M., 1994. Reductive dehalogenation of hexachloroethane, carbon tetrachloride, and bromoform by anthrahydroquinone disulfonate and humic acid. Environmental Science Technology 28, 2393-2401.

DeLaune, R.D., Hambrick III, G.A., Patrick Jr., W.H., 1980. Degradation of hydrocarbons in oxidized and reduced sediments. Marine Pollution Bulletin 11, 103-106.

Derby, H.A., Hammer, B.W., 1931. Bacteriology of butter. IV Bacteriological studies in surface taint butter. Research Bulletin of the Iowa Agricultural Experiment Station.

Dobbin, P.S., Powell, A.K., McEwan, A.G., Richardson, D.J., 1995. The influence of chelating agents upon the dissimilatory reduction of Fe(III) by Shewanella putrefaciens. Biometals 8, 163-173.

Edwards, E.A., Grbic-Galic, D., 1994. Anaerobic degradation of toluene and o-xylene by a methanogenic consortium. Applied and Environmental Microbiology 60, 313-322.

Ehlers, G.A.C., Forrester, S.T., Scherr, K.E., Loibner, A.P., Janik, L.J., 2010. Influence of the nature of soil organic matter on the sorption behaviour of pentadecane as determined by PLS analysis of mid-infrared DRIFT and solid-state ^{13}C NMR spectra. Environmental Pollution 158, 285-291.

Ehlers, G.A.C., Loibner, A.P., 2006. Linking organic pollutant (bio)availability with geosorbent properties and biomimetic methodology: A review of geosorbent characterisation and (bio)availability prediction. Environmental Pollution 141, 494-512.

Evans, P.J., 2000. A Novel Ferric Iron Bioavailability Assay, in: Wickramanayake, G.B., al., e. (Eds.), Risk, Regulatory, and Monitoring Considerations. Battelle Press,, Columbus, OH, USA, pp. 167-174.

Fennessey, C.M., Jones, M.E., Taillefert, M., DiChristina, T.J., 2010. Siderophores Are Not Involved in Fe(III) Solubilization during Anaerobic Fe(III) Respiration by Shewanella oneidensis MR-1. Applied and Environmental Microbiology 76, 2425-2432.

Field, J.A., Cervantes, F.J., Van der Zee, F.P., Lettinga, G., 2000. Role of quinones in the biodegradation of priority pollutants: A review. Water Science and Technology 42, 215-222.

Finneran, K.T., Lovley, D.R., 2001. Anaerobic degradation of methyl tert-butyl ether (MTBE) and tert-butyl alcohol (TBA). Environmental Science and Technology 35, 1785-1790.

Friedl, G., Wehrli, B., Manceau, A., 1997. Solid phases in the cycling of manganese in eutrophic lakes: New insights from EXAFS spectroscopy. Geochimica et Cosmochimica Acta 61, 275-290.

Fultz, M.L., Durst, R.A., 1982. Mediator compounds for the electrochemical study of biological redox systems: a compilation. Analytica Chimica Acta 140, 1-18.

Gorby, Y.A., Yanina, S., McLean, J.S., Rosso, K.M., Moyles, D., Dohnalkova, A., Beveridge, T.J., Chang, I.S., Kim, B.H., Kim, K.S., Culley, D.E., Reed, S.B., Romine, M.F., Saffarini, D.A., Hill, E.A., Shi, L., Elias, D.A., Kennedy, D.W., Pinchuk, G., Watanabe, K., Ishii, S., Logan, B., Nealson, K.H., Fredrickson, J.K., 2006. Electrically conductive bacterial nanowires produced by Shewanella oneidensis strain MR-1 and other microorganisms. Proceedings of the National Academy of Sciences of the United States of America 103, 11358-11363.

Gray, H.B., Winkler, J.R., 1996. Electron Transfer in Proteins. Annual Review of Biochemistry 65, 537-561.

Greene, A.C., Patel, B.K.C., Sheehy, A.J., 1997. Deferribacter thermophilus gen. nov., sp. nov., a novel thermophilic manganese- and iron-reducing bacterium isolated from a petroleum reservoir. International Journal of Systematic Bacteriology 47, 505-509.

Greene, A.C., Patel, B.K.C., Yacob, S., 2009. Geoalkalibacter subterraneus sp. nov., an anaerobic Fe(III)- and Mn(IV)-reducing bacterium from a petroleum reservoir, and emended descriptions of the family Desulfuromonadaceae and the genus Geoalkalibacter. International Journal of Systematic and Evolutionary Microbiology 59, 781-785.

Gu, B., Bian, Y., Miller, C.L., Dong, W., Jiang, X., Liang, L., 2011. Mercury reduction and complexation by natural organic matter in anoxic environments. Proceedings of the National Academy of Sciences 108, 1479-1483.

Guo, J., Liu, H., Qu, J., Lian, J., Zhao, L., Jefferson, W., Yang, J., 2012. The structure activity relationship of non-dissolved redox mediators during azo dye bio-decolorization processes. Bioresource Technology 112, 350-354.

Haas, J.R., Dichristina, T.J., 2002. Effects of FE(III) chemical speciation on dissimilatory FE(III) reduction by shewanella putrefaciens. Environmental Science and Technology 36, 373-380.

Hasinger, M., Scherr, Kerstin E., Lundaa, T., Bräuer, L., Zach, C., Loibner, A.P., 2012. Changes in *iso-* and *n*-alkane distribution during biodegradation of crude oil under nitrate and sulphate reducing conditions. Journal of Biotechnology 157, 490-498.

Heider, J., Spormann, A.M., Beller, H.R., Widdel, F., 1998. Anaerobic bacterial metabolism of hydrocarbons. FEMS Microbiology Reviews 22, 459-473.

Hernandez, M.E., Kappler, A., Newman, D.K., 2004. Phenazines and Other Redox-Active Antibiotics Promote Microbial Mineral Reduction. Applied and Environmental Microbiology 70, 921-928.

Hernandez, M.E., Newman, D.K., 2001. Extracellular electron transfer. Cellular and Molecular Life Sciences 58, 1562-1571.

Hirano, H., Yoshida, T., Fuse, H., Endo, T., Habe, H., Nojiri, H., Omori, T., 2003. Marinobacterium sp. strain DMS-S1 uses dimethyl sulphide as a sulphur source after light-dependent transformation by excreted flavins. Environmental Microbiology 5, 503-509.

Istok, J.D., Senko, J.M., Krumholz, L.R., Watson, D., Bogle, M.A., Peacock, A., Chang, Y.J., White, D.C., 2004. In situ bioreduction of technetium and uranium in a nitrate-contaminated aquifer. Environmental Science and Technology 38, 468-475.

Jiang, J., Kappler, A., 2008. Kinetics of microbial and chemical reduction of humic substances: Implications for electron shuttling. Environmental Science and Technology 42, 3563-3569.

Jiao, Y., Qian, F., Li, Y., Wang, G., Saltikov, C.W., Gralnick, J.A., 2011. Deciphering the electron transport pathway for graphene oxide reduction by Shewanella oneidensis MR-1. Journal of Bacteriology 193, 3662-3665.

Jones, J.G., 1977. The effect of environmental factors on estimated viable and total populations of planktonic bacteria in lakes and experimental enclosures. Freshwater Biology 7, 67-91.

Kang, S.-H., Choi, W., 2009. Oxidative degradation of organic compounds using zero-valent iron in the presence of natural organic matter serving as an electron shuttle. Environmental Science and Technology 43, 818-883.

Karickhoff, S.W., Brown, D.S., Scott, T.A., 1979. Sorption of hydrophobic pollutants on natural sediments. Water Research 13, 241-248.

Kashefi, K., Tor, J.M., Nevin, K.P., Lovley, D.R., 2001. Reductive Precipitation of Gold by Dissimilatory Fe(III)-Reducing Bacteria and Archaea. Applied and Environmental Microbiology 67, 3275-3279.

Kerisit, S., Rosso, K.M., Dupuis, M., Valiev, M., 2007. Molecular Computational Investigation of Electron-Transfer Kinetics Across Cytochrome–Iron Oxide Interfaces. The Journal of Physical Chemistry C 111, 11363-11375.

Kouzuma, A., Hashimoto, K., Watanabe, K., 2012. Roles of siderophore in manganese-oxide reduction by Shewanella oneidensis MR-1. FEMS Microbiology Letters 326, 91-98.

Kumagai, Y., Shinkai, Y., Miura, T., Cho, A.K., 2012. The chemical biology of naphthoquinones and its environmental implications, pp. 221-247.

Kunapuli, U., Griebler, C., Beller, H.R., Meckenstock, R.U., 2008. Identification of intermediates formed during anaerobic benzene degradation by an iron-reducing enrichment culture. Environmental Microbiology 10, 1703-1712.

Kwon, M.J., Finneran, K.T., 2006. Microbially mediated biodegradation of hexahydro-1,3,5-trinitro-1,3,5- triazine by extracellular electron shuttling compounds. Applied and Environmental Microbiology 72, 5933-5941.

Kwon, M.J., Finneran, K.T., 2008. Biotransformation products and mineralization potential for hexahydro-1,3,5-trinitro-1,3,5-triazine (RDX) in abiotic versus biological degradation pathways with anthraquinone-2,6-disulfonate (AQDS) and Geobacter metallireducens. Biodegradation 19, 705-715.

Langenhoff, A.A.M., Brouwers-Ceiler, D.L., Engelberting, J.H.L., Quist, J.J., Wolkenfelt, J.G.P.N., Zehnder, A.J.B., Schraa, G., 1997. Microbial reduction of manganese coupled to toluene oxidation. FEMS Microbiology Ecology 22, 119-127.

Lee, C.H., Lewis, T.A., Paszczynski, A., Crawford, R.L., 1999. Identification of an extracellular catalyst of carbon tetrachloride dehalogenation from Pseudomonas stutzeri strain KC as pyridine-2,6-bis(thiocarboxylate). Biochemical and Biophysical Research Communications 261, 562-566.

Lee, K.Y., Bosch, J., Meckenstock, R.U., 2012. Use of metal-reducing bacteria for bioremediation of soil contaminated with mixed organic and inorganic pollutants. Environmental Geochemistry and Health 34, 135-142.

Lee, W., Batchelor, B., 2002. Abiotic Reductive Dechlorination of Chlorinated Ethylenes by Iron-Bearing Soil Minerals. 1. Pyrite and Magnetite. Environmental Science & Technology 36, 5147-5154.

Li, Y., Yu, S., Strong, J., Wang, H., 2012. Are the biogeochemical cycles of carbon, nitrogen, sulfur, and phosphorus driven by the "Fe III-Fe II redox wheel" in dynamic redox environments? Journal of Soils and Sediments 12, 683-693.

Lies, D.P., Hernandez, M.E., Kappler, A., Mielke, R.E., Gralnick, J.A., Newman, D.K., 2005. Shewanella oneidensis MR-1 Uses Overlapping Pathways for Iron Reduction at a Distance and by Direct Contact under Conditions Relevant for Biofilms. Applied and Environmental Microbiology 71, 4414-4426.

Lim, M.S., Yeo, I.W., Roh, Y., Lee, K.K., Jung, M.C., 2008. Arsenic reduction and precipitation by shewanella sp.: Batch and column tests. Geosciences Journal 12, 151-157.

Lloyd, J.R., 2003. Microbial reduction of metals and radionuclides. FEMS Microbiology Reviews 27, 411-425.

Logan, B.E., 2009. Exoelectrogenic bacteria that power microbial fuel cells. Nature Reviews Microbiology 7, 375-381.

Loibner, A.P., Szolar, O.H.J., Braun, R., Hirmann, D., 2004. Toxicity testing of 16 priority polycyclic aromatic hydrocarbons using Lumistox®. Environmental Toxicology and Chemistry 23, 557-564.

Lovley, D.R., 2011. Live wires: direct extracellular electron exchange for bioenergy and the bioremediation of energy-related contamination. Energy & Environmental Science 4, 4896-4906.

Lovley, D.R., Baedecker, M.J., Lonergan, D.J., Cozzarelli, I.M., Phillips, E.J.P., Siegel, D.I., 1989. Oxidation of aromatic contaminants coupled to microbial iron reduction. Nature 339, 297-300.

Lovley, D.R., Coates, J.D., Blunt-Harris, E.L., Phillips, E.J.P., Woodward, J.C., 1996a. Humic substances as electron acceptors for microbial respiration. Nature 382, 445-448.

Lovley, D.R., Fraga, J.L., Blunt-Harris, E.L., Hayes, L.A., Phillips, E.J.P., Coates, J.D., 1998. Humic substances as a mediator for microbially catalyzed metal reduction. Acta Hydrochimica et Hydrobiologica 26, 152-157.

Lovley, D.R., Fraga, J.L., Coates, J.D., Blunt-Harris, E.L., 1999. Humics as an electron donor for anaerobic respiration. Environmental Microbiology 1, 89-98.

Lovley, D.R., Giovannoni, S.J., White, D.C., Champine, J.E., Phillips, E.J.P., Gorby, Y.A., Goodwin, S., 1993. Geobacter metallireducens gen. nov. sp. nov., a microorganism capable of coupling the complete oxidation of organic compounds to the reduction of iron and other metals. Archives of Microbiology 159.

Lovley, D.R., Holmes, D.E., Nevin, K.P., 2004. Dissimilatory Fe(III) and Mn(IV) reduction. Advances in Microbial Physiology 49, 219-286.

Lovley, D.R., Phillips, E.J., 1992. Reduction of uranium by Desulfovibrio desulfuricans. Applied and Environmental Microbiology 58, 850-856.

Lovley, D.R., Phillips, E.J., Lonergan, D.J., Widman, P.K., 1995. Fe(III) and S0 reduction by Pelobacter carbinolicus. Applied and Environmental Microbiology 61, 2132-2138.

Lovley, D.R., Phillips, E.J.P., Gorby, Y.A., Landa, E.R., 1991. Microbial reduction of uranium. Nature 350, 413-416.

Lovley, D.R., Stolz, J.F., Nord, G.L., Phillips, E.J.P., 1987. Anaerobic production of magnetite by a dissimilatory iron-reducing microorganism. Nature 330, 252-254.

Lovley, D.R., Ueki, T., Zhang, T., Malvankar, N.S., Shrestha, P.M., Flanagan, K.A., Aklujkar, M., Butler, J.E., Giloteaux, L., Rotaru, A.E., Holmes, D.E., Franks, A.E., Orellana, R., Risso, C., Nevin, K.P., 2011. Geobacter. The Microbe Electric's Physiology, Ecology, and Practical Applications, pp. 1-100.

Lovley, D.R., Woodward, J.C., Chapelle, F.H., 1996b. Rapid anaerobic benzene oxidation with a variety of chelated Fe(III) forms. Applied and Environmental Microbiology 62, 288-291.

Lown, W.J., 1983. The mechanism of action of quinone antibiotics. Molecular and Cellular Biochemistry 55.

Marsili, E., Baron, D.B., Shikhare, I.D., Coursolle, D., Gralnick, J.A., Bond, D.R., 2008. Shewanella secretes flavins that mediate extracellular electron transfer. Proceedings of the National Academy of Sciences of the United States of America 105, 3968-3973.

Martin, J.-M., Meybeck, M., 1979. Elemental mass-balance of material carried by major world rivers. Marine Chemistry 7, 173-206.

Maymo-Gatell, X., Anguish, T., Zinder, S.H., 1999. Reductive dechlorination of chlorinated ethenes and 1,2-dichloroethane by 'Dehalococcoides ethenogenes' 195. Applied and Environmental Microbiology 65, 3108-3113.

McCarthy, J.F., Jimenez, B.D., 1985. Interactions between polycyclic aromatic hydrocarbons and dissolved humic material: Binding and dissociation. Environmental Science and Technology 19, 1072-1076.

Meckstroth, M.L., Norris, B.J., Heineman, W.R., 1981. 387 — Mediator-titrants for thin-layer spectroelectrochemical measurement of biocomponent Uo′ and n values. Bioelectrochemistry and Bioenergetics 8, 63-70.

Michaelis, L., Schubert, M.P., 1938. The Theory of Reversible Two-step Oxidation Involving Free Radicals. Chemical Reviews 22, 437-470.

Morel, F.M.M., Hering, J.G., 1993. Principles and applications of aquatic chemistry. Wiley Interscience, Ney York, NY.

Muller, F., Grande, H.J., Harding, L.J., Dunham, W.R., Visser, A.J., Reinders, J.H., Hemmerich, P., Ehrenberg, A., 1981. On the molecular and submolecular structure of the semiquinone cations of alloxazines and isoalloxazines as revealed by electron-paramagnetic-resonance spectroscopy. European Journal of Biochemistry 116, 17-25.

Myers, C.R., Myers, J.M., 1992. Localization of cytochromes to the outer membrane of anaerobically grown Shewanella putrefaciens MR-1. Journal of Bacteriology 174, 3429-3438.

Myers, C.R., Nealson, K.H., 1988. Bacterial manganese reduction and growth with manganese oxide as the sole electron acceptor. Science Science 240, 1319-1321.

Neilands, J.B., 1995. Siderophores: Structure and Function of Microbial Iron Transport Compounds. Journal of Biological Chemistry 270, 26723-26726.

Nevin, K.P., Lovley, D.R., 2000. Potential for Nonenzymatic Reduction of Fe(III) via Electron Shuttling in Subsurface Sediments. Environmental Science & Technology 34, 2472-2478.

Nevin, K.P., Lovley, D.R., 2002a. Mechanisms for Accessing Insoluble Fe(III) Oxide during Dissimilatory Fe(III) Reduction by Geothrix fermentans. Applied and Environmental Microbiology 68, 2294-2299.

Nevin, K.P., Lovley, D.R., 2002b. Mechanisms for Fe(III) Oxide Reduction in Sedimentary Environments. Geomicrobiology Journal 19, 141-159.

Newman, D.K., Kolter, R., 2000. A role for excreted quinones in extracellular electron transfer. Nature 405, 94-97.

Nguyen, T.T., Janik, L.J., Raupach, M., 1991. Diffuse reflectance infrared Fourier transform (DRIFT) spectroscopy in soil studies. Australian Journal of Soil Research 29, 49-67.

O'Loughlin, E.J., 2008. Effects of Electron Transfer Mediators on the Bioreduction of Lepidocrocite (γ-FeOOH) by Shewanella putrefaciens CN32. Environmental Science & Technology 42, 6876-6882.

Perminova, I., Hatfield, K., 2005. Remediation Chemistry of Humic Substances: Theory and Implications for Technology, in: Perminova, I., Hatfield, K., Hertkorn, N. (Eds.), Use of Humic Substances to Remediate Polluted Environments: From Theory to Practice. Springer, Dordrecht, The Netherlands.

Perminova, I.V., Grechishcheva, N.Y., Petrosyan, V.S., 1999. Relationships between structure and binding affinity of humic substances for polycyclic aromatic hydrocarbons: Relevance of molecular descriptors. Environmental Science and Technology 33, 3781-3787.

Perminova, I.V., Kovalenko, A.N., Schmitt-Kopplin, P., Hatfield, K., Hertkorn, N., Belyaeva, E.Y., Petrosyan, V.S., 2005. Design of quinonoid-enriched humic materials with enhanced redox properties. Environmental Science and Technology 39, 8518-8524.

Pham, T.H., Boon, N., De Maeyer, K., Höfte, M., Rabaey, K., Verstraete, W., 2008. Use of Pseudomonas species producing phenazine-based metabolites in the anodes of microbial fuel cells to improve electricity generation. Applied Microbiology and Biotechnology 80, 985-993.

Phillips, E.J., Lovley, D.R., Roden, E.E., 1993. Composition of Non-Microbially Reducible Fe(III) in Aquatic Sediments. Applied and Environmental Microbiology 59, 2727-2729.

Ratasuk, N., Nanny, M.A., 2007. Characterization and Quantification of Reversible Redox Sites in Humic Substances. Environmental Science & Technology 41, 7844-7850.

Reguera, G., McCarthy, K.D., Mehta, T., Nicoll, J.S., Tuominen, M.T., Lovley, D.R., 2005. Extracellular electron transfer via microbial nanowires. Nature 435, 1098-1101.

Reimann, C., de Caritat, P., 2012. New soil composition data for Europe and Australia: Demonstrating comparability, identifying continental-scale processes and learning lessons for global geochemical mapping. Science of The Total Environment 416, 239-252.

Richter, K., Schicklberger, M., Gescher, J., 2012. Dissimilatory Reduction of Extracellular Electron Acceptors in Anaerobic Respiration. Applied and Environmental Microbiology 78, 913-921.

Rooney-Varga, J.N., Anderson, R.T., Fraga, J.L., Ringelberg, D., Lovley, D.R., 1999. Microbial communities associated with anaerobic benzene degradation in a petroleum-contaminated aquifer. Applied and Environmental Microbiology 65, 3056-3063.

Royer, R.A., Burgos, W.D., Fisher, A.S., Unz, R.F., Dempsey, B.A., 2002. Enhancement of biological reduction of hematite by electron shuttling and Fe(II) complexation. Environmental Science and Technology 36, 1939-1946.

Scheffer, F., Schachtschabel, P., 2002. Lehrbuch der Bodenkunde. Spektrum Akademischer Verlag.

Scherr, K.E., Nahold, M., Lantschbauer, W., Loibner, A.P., 2011. Sequential application of electron donors and humic acids for the anaerobic bioremediation of chlorinated aliphatic hydrocarbons. New Biotechnology 1, 116-125.

Schink, B., 1997. Energetics of syntrophic cooperation in methanogenic degradation. Microbiology and Molecular Biology Reviews 61, 262-280.

Schnitzer, M., 1989. Humic Substances: Chemistry and Reactions, in: Schnitzer, M., Khan, S.U. (Eds.), Soil Organic Matter. Elsevier Science Publishing Company, New York, NY.

Scholz, D., Kucklick, J., Hayward Walker, A., Pavia, R., 1998. Managing Spills of Oil and Chemical Materials. NOAA's State of the Coast Report. National Oceanic and Atmospheric Administration (NOAA), Silver Spring, MD, US.

Scott, D.T., McKnight, D.M., Blunt-Harris, E.L., Kolesar, S.E., Lovley, D.R., 1998. Quinone moieties act as electron acceptors in the reduction of humic substances by humics-reducing microorganisms. Environmental Science and Technology 32, 2984-2989.

Shang, Y., Chen, C., Li, Y., Zhao, J., Zhu, T., 2012. Hydroxyl radical generation mechanism during the redox cycling process of 1,4-naphthoquinone. Environmental Science and Technology 46, 2935-2942.

Shi, L., Squier, T.C., Zachara, J.M., Fredrickson, J.K., 2007. Respiration of metal (hydr)oxides by Shewanella and Geobacter: a key role for multihaem c-type cytochromes. Molecular Microbiology 65, 12-20.

Siegert, M., Cichocka, D., Herrmann, S., Gründger, F., Feisthauer, S., Richnow, H.H., Springael, D., Krüger, M., 2011. Accelerated methanogenesis from aliphatic and aromatic hydrocarbons under iron- and sulfate-reducing conditions. FEMS Microbiology Letters 315, 6-16.

Snoeyenbos-West, O.L., Nevin, K.P., Anderson, R.T., Lovley, D.R., 2000. Enrichment of Geobacter Species in Response to Stimulation of Fe(III) Reduction in Sandy Aquifer Sediments. Microbial Ecology 39, 153-167.

Sposito, G., 2011. Electron Shuttling by Natural Organic Matter: Twenty Years After, in: Tratnyek, P. (Ed.), Aquatic Redox Chemistry. Americal Chemical Society, Washington, D.C., pp. 113-127.

Struyk, Z., Sposito, G., 2001. Redox properties of standard humic acids. Geoderma 102, 329-346.

Taillefert, M., Beckler, J.S., Carey, E., Burns, J.L., Fennessey, C.M., DiChristina, T.J., 2007. Shewanella putrefaciens produces an Fe(III)-solubilizing organic ligand during anaerobic respiration on insoluble Fe(III) oxides. Journal of Inorganic Biochemistry 101, 1760-1767.

Thamdrup, B., 2000. Bacterial Manganese and Iron Reduction in Aquatic Sediments, in: Schink, B. (Ed.), Advances in Microbial Ecology. Kluwer Academic / Plenum Publishers, New York, NY, pp. 41-84.

Torsvik, V., Goksoyr, J., Daae, F.L., 1990. High diversity in DNA of soil bacteria. Applied and Environmental Microbiology 56, 782-787.

Tratnyek, P.G., Macalady, D.L., 1989. Abiotic reduction of nitro aromatic pesticides in anaerobic laboratory systems. Journal of Agricultural and Food Chemistry 37, 248-254.

Turick, C.E., Tisa, L.S., Caccavo, F., Jr., 2002. Melanin production and use as a soluble electron shuttle for Fe(III) oxide reduction and as a terminal electron acceptor by Shewanella algae BrY. Applied and Environmental Microbiology 68, 2436-2444.

US-EPA, 2012. SPCC Rule I Emergency Management I US EPA. OSWER, Office of Emergency Management.

Van der Zee, F.P., Cervantes, F.J., 2009. Impact and application of electron shuttles on the redox (bio)transformation of contaminants: A review. Biotechnology Advances 27, 256-277.

Venkateswaran, K., Moser, D.P., Dollhopf, M.E., Lies, D.P., Saffarini, D.A., MacGregor, B.J., Ringelberg, D.B., White, D.C., Nishijima, M., Sano, H., Burghardt, J., Stackebrandt, E., Nealson, K.H., 1999. Polyphasic taxonomy of the genus Shewanella and description of Shewanella oneidensis sp. nov. International Journal of Systematic Bacteriology 49, 705-724.

von Canstein, H., Ogawa, J., Shimizu, S., Lloyd, J.R., 2008. Secretion of Flavins by Shewanella Species and Their Role in Extracellular Electron Transfer. Applied and Environmental Microbiology 74, 615-623.

Wang, Y., Newman, D.K., 2008. Redox reactions of phenazine antibiotics with ferric (hydr)oxides and molecular oxygen. Environmental Science and Technology 42, 2380-2386.

Watanabe, K., Manefiled, M., Lee, M., Kouzuma, A., 2009. Electron shuttles in biotechnology. Current Opinion in Biotechnology: 20, 633-641.

Weber, K.A., Achenbach, L.A., Coates, J.D., 2006. Microorganisms pumping iron: anaerobic microbial iron oxidation and reduction. Nature Reviews Microbiology 4, 752-764.

Widdel, F., Rabus, R., 2001. Anaerobic biodegradation of saturated and aromatic hydrocarbons. Current Opinion in Biotechnology 12, 259-276.

Wolf, M., Kappler, A., Jiang, J., Meckenstock, R.U., 2009. Effects of Humic Substances and Quinones at Low Concentrations on Ferrihydrite Reduction by Geobacter metallireducens. Environmental Science and Technology 43, 5679-5685.

Yamazaki, S., Kano, K., Ikeda, T., Isawa, K., Kaneko, T., 1999. Role of 2-amino-3-carboxy-1,4-naphthoquinone, a strong growth stimulator for bifidobacteria, as an electron transfer mediator for NAD(P)(+) regeneration in Bifidobacterium longum. Biochimica et Biophysica Acta 1428, 241-250.

Zheng, W., Liang, L., Gu, B., 2012. Mercury Reduction and Oxidation by Reduced Natural Organic Matter in Anoxic Environments. Environmental Science & Technology 46, 292-299.

Zhou, M., Luo, H., Li, Z., Wu, F., Huang, C., Ding, Z., Li, R., 2012. Recent advances in screening of natural products for antimicrobial agents. Combinatorial Chemistry and High Throughput Screening 15, 306-315.

Zhu, M., Farrow, C.L., Post, J.E., Livi, K.J.T., Billinge, S.J.L., Ginder-Vogel, M., Sparks, D.L., 2012. Structural study of biotic and abiotic poorly-crystalline manganese oxides using atomic pair distribution function analysis. Geochimica et Cosmochimica Acta 81, 39-55.

Chapter 9

Microbial Techniques for Hydrocarbon Exploration

M.A. Rasheed, D.J. Patil and A.M. Dayal

Additional information is available at the end of the chapter

http://dx.doi.org/10.5772/50885

1. Introduction

Bacteria are ubiquitous in distribution and their exceptionally high adaptability to grow on different nutrient sources form the basis of microbial prospecting. Several investigators have used bacteria that degrade hydrocarbons, used as indicators for finding oil and gas reservoirs. Microbial prospecting method for hydrocarbons is a surface exploration method based on the premise that the light gaseous hydrocarbons namely methane (C_1), ethane (C_2), propane (C_3) and butane (C_4) migrate upward from subsurface petroleum accumulations by diffusion and effusion (Horvitz, 1939) and are utilized by a variety of microorganisms present in the sub-soil ecosystem. The methane, ethane, propane, and butane-oxidizing bacteria exclusively use these gases as a carbon source for their metabolic activities and growth. These bacteria are mostly found enriched in the shallow soils/sediments above hydrocarbon bearing structures and can differentiate between hydrocarbon prospective and non-prospective areas (Tucker and Hitzman, 1994). The detection of various groups of methane, ethane, propane or butane oxidizing bacteria, in the surface soils or sediments, helps to evaluate the prospects for hydrocarbon exploration. Microbial prospecting is a surface prospecting technique,, well known in the realm of hydrocarbon exploration researchers. Microbial anomalies have been proved to be reliable indicators of oil and gas in the sub-surface (Pareja, 1994). The direct and positive relationship between the microbial population and the hydrocarbon concentration in the soils have been observed in various producing reservoirs worldwide (Miller, 1976; Sealy 1974a, 1974b; Wagner et al., 2002). These light hydrocarbon are utilized by a phylogenetically diverse group of bacteria belonging to genera of *Brevibacterium, Corynebacterium, Flavobacterium, Mycobacterium, Nocardia, Pseudomonas, Rhodococcus* etc., (Perry and Williams, 1968; Vestal and Perry, 1971). The microbial prospecting method involves the collection of sub-soil samples from the study/survey area, packing of samples, preservation and storage of samples in pre-sterilized sample bags under aseptic and cold conditions till analysis and followed by isolation and enumeration of hydrocarbon utilizing bacteria. The contour maps for population density of hydrocarbon oxidizing bacteria are drawn and the data is integrated with other geo-scientific and geophysical data to find the hydrocarbon prospectivity of the area.

The microbial survey was first proposed and applied in the U.S.S.R. Early use of bacterial soil flora as a means of detecting gas seepages was a development stemming from soil gas surveys performed in the U.S.S.R (Mogilevskii 1959). The initial microbial investigations of Mogilevskii and his associates in the field of petroleum prospecting incited the interest of petroleum geochemists worldwide. The microbial surveys were first proposed and applied in the U.S.S.R, (Mogilevski, 1940). It has been shown that out of 20 microbial anomalies, 16 were proved by successful drilling. These investigations initiated the interest of petroleum geochemists worldwide. Sealy (1974a) in USA, carried out microbial prospecting surveys and showed a positive correlation of 85.7%. Miller (1976) reported microbial survey carried out in the oil fields of the U.S.A, in which the microbial activity profile indicated a good contrast between oil fields and nearby dry area. Beghtel (1987) predicted hydrocarbon potential of 18 wildcat wells in the Kansas; out of which, 13 have proved to be commercial producers of oil and gas. As a result, microbial methods for detecting petroleum gas in the soil and waters have been tested in Europe, U.S.A and elsewhere. A sustained effort for the purpose of efficiently utilizing microbiological prospecting methods had been underway since 1938 including independent surveys, surveys coordinated with geochemical and geophysical operation, microbiological tests of both soils and sub surface waters and analysis for methane oxidizing as well as higher hydrocarbon oxidizing bacteria. However, oil and gas fields also build up micro-seepages at the sub-surface soils, and these micro-seepages are detectable using a variety of analytical techniques that have been developed in the past 70 years. Geochemists developed the basis for these new surface prospecting methods. The pioneers were Laubmeyer (1933) in Germany, Rosaire (1939) and Horvitz (1939) in the United States. Using methods such as extraction of adsorbed hydrocarbon gases from surface samples, they documented a correlation between higher hydrocarbon concentrations and oil and gas fields. At almost at the same time, the microbiologists Mogilewskii (1938, 1940) in the U.S.S.R. and Taggart (1941) and Blau (1942) in the United States described the use of hydrocarbon-oxidizing bacteria (HCO), when measured in surface soil samples, as an indicator for oil and gas fields in the deeper subsurface. In the 1950s and early 1960s, many relevant publications came from the United States (Updegraff et al., 1954; Davis, 1956). The U.S.S.R. (Bokova et al., 1947; Davis, 1967 and Sealy, 1974 a, b) have published reviews of this early work. Several microbiological methods for detecting the distribution and activity of HCO were developed, such as enumeration of cell content in soil samples, measuring gas-consumption rates and radioautography. The field trials conducted by Sealy (1974) in U.S.A., using microbiological techniques, showed that out of 89 locations tested in west Texas, wildcats predicated as productive and non-productive correlation of 54% and 92% respectively. Sealy (1974) reported positive prognosis for producers, 8 out of 11 and for dry holes 28 out of 31, having an overall correct prognosis of 85.7%. Miller (1976) reported microbial survey of a number of oil fields in the U.S.A. some of which were primarily stratigraphic and showed that microbial activity profiles indicated a good contrast between oil fields and nearby dry areas. Beghtel et al., (1987) described a new Microbial Oil Survey Technique, named MOST, which uses the higher butanol resistance of butane-oxidizing bacteria to detect hydrocarbon micro-seepages. Microbial anomalies have been proved to be reliable indicators of oil and gas in the subsurface (Pareja et al., 1994). Hydrocarbon micro-seepage detection adds value to 2-D and 3-D seismic by identifying those features that are charged with hydrocarbons (Schumacher, 1997). There is a direct positive relationship between the increased hydrocarbon concentrations and increased hydrocarbon indicating microbial populations. This relationship is easily measurable and distinctly reproducible. Microbial anomalies can also used for development of field and reservoir characterization studies (Hitzman et al 1999, Hitzman 1994, Schumacher et al 1997).

Microbial Prospecting survey has been widely used in Germany since 1961 and a total of 17 oil and gas fields were identified. The success rate of Microbial Prospecting for Oil and Gas (MPOG) method has been reported to be 90%. This method can be integrated with geological, geochemical and geophysical methods to evaluate the hydrocarbon prospect of an area and to prioritize the drilling locations; thereby, reducing drilling risks and achieving higher success in petroleum exploration (Wagner et al., 2002).

1.1. Oxidation reduction zones

Bacteria and other microbes play a profound role in the oxidation of migrating hydrocarbons. Their activities are directly or indirectly responsible for many of the diverse surface manifestations of petroleum seepage. These activities, coupled with long-term migration of hydrocarbons, lead to the development of near-surface oxidation-reduction zones that favor the formation of this variety of hydrocarbon-induced chemical and mineralogical changes. This seep-induced alteration is highly complex and its varied surface expressions have led to the development of an equally varied number of geochemical exploration techniques. Some detect hydrocarbons directly in surface and seafloor samples, others detect seep-related microbial activity, and still others measure the secondary effects of hydrocarbon-induced alteration.

The activities of hydrocarbon oxidizing bacteria cause the development of near-surface oxidation –reduction zones and the alteration of soils and sediments above the reservoirs. These changes form the basis for other surface exploration techniques, such as soil carbonate, magnetic, electrical, radioactivity and satellite-based methods (Richers et al., 1982; Schumacher, 1996).

2. Indicators of microbial prospecting

Methane, ethane, propane and butane oxidizing bacteria have been used by various researchers as indicator microbes for prospecting of oil and gas.

3. Methane oxidizing bacteria

Methane oxidizing bacteria were the first type of bacteria studied to identify the location of petroleum accumulations. The presence of methane oxidizing bacteria in the soil, in the absence of cellulose oxidizing bacteria, has been interpreted as indicating the presence of methane exhalation from the subsurface (Tedesco, 1995). Mogilewskii (1938) described the possibility of using methane-oxidizing bacteria for gas exploration. Methane oxidizing bacteria were found in the petroleum prospecting operations of the Soviet Union (Kartsev et al., 1959). Methane oxidizing bacteria have been isolated from soil as a means of prospecting from natural gas and/or oil deposits (Brisbane and Ladd, 1965). The methane oxidizing bacteria are usually predominant over gas fields as the gas reservoirs are commonly dominated by methane. Microbial Prospection for Oil and Gas (MPOG) method establish the separate activities of methane-oxidizing bacteria as gas indicators and those bacteria that oxidize only ethane and higher hydrocarbons as oil indicators, it is possible to differentiate between oil fields with and without free gas cap, and gas fields (Wagner et al., 2002). Methane oxidizing bacteria have been deployed as one of the indicator microbes (Jain et al., 1991). Thermogenic processes produce methane and substantial amounts of other saturated hydrocarbons by irreversible reaction of residual organic matter or kerogen (Klusman, 1993). Whittenbury et al., (1970) reported the isolation of more than 100 strains of methane-oxidizing bacteria. Hanson and Wattenberg (1991) and Hanson and Hanson (1996) gave overviews of the ecology of methylotrophic bacteria and their role in the methane cycle. The methane oxidizing bacteria (Methylotrophs) are *Methylococcus, Methylomonas, Methylobacter, Methylocyctis, Methylosinus, Methylobacterium* etc., (Hanson and Hanson 1996). However, methane oxidizing bacteria are known to be poor indicators in petroleum prospecting because methane can occur in the absence of petroleum deposits and moreover, some methane oxidizing bacteria are unable to oxidize other aliphatic hydrocarbons. Detection of ethane and longer chain hydrocarbon oxidizing bacteria on the other hand provides presumptive evidence for a hydrocarbon seep and an underlying petroleum reservoir (Davis, 1967). The principal advantage of using methane-oxidizing bacteria for petroleum prospecting is the preponderance of methane in petroleum gas and it is the most mobile of petroleum hydrocarbons and this cannot be ignored.

3.1. Microbial oxidation of methane

Microbial methane oxidation starts with an activation of methane by methane-oxygenase (MMO), this enzyme is a mono-oxygenase because only one atom of the dioxygen molecule is inserted into the methane molecule to produce methanol; and leads, in the presence of oxygen, to methanol and then to formaldehyde. Formaldehyde can be directly assimilated to produce biomass or oxidized to CO_2 for the generation of energy (Leadbetter and Forster 1958). Because of the high specialization of these bacteria, methane oxidizers can be isolated from all other bacteria. The successful detection of these specialists indicates methane occurrence in soil samples (Wagner et al., 2002).

4. Ethane, propane and butane oxidizing bacteria

The Soviets were first to use microbial prospecting method. They established that certain bacteria specifically consume ethane, propane, or butane, but not methane. These hydrocarbons are assumed to be from petroleum migrating from depth and are not associated with generation in the soil (Tedesco 1995). The isolation and enumeration of specific C_{2+} alkane-oxidizing bacteria have been used as indirect petroleum prospecting method (Davis, 1967, Wagner et al., 2002). The detection of bacteria that oxidize *n*-alkanes having chain lengths of 2 to 8 carbon atoms, without any adaptation period, indicates the existence of short-chain hydrocarbons in the investigated soil samples, and thus can indicate the presence of oil accumulations in the subsurface. In those areas in which both short-chain hydrocarbons and methane are detected in this manner, one can assume a thermogenic gas with significant quantities of short-chain hydrocarbons, depending on the intensity of the signals (Wagner et al., 2002). The short-chain hydrocarbons ethane, propane, and butane can be used by a large number of bacteria (*Mycobacteria, Flavobacteria, Nocardia,* and *Pseudomonas*). The microbial anomalies have proven to be reliable indicators of oil and gas in the subsurface (Beghtel et al, 1987; Lopez et al., 1993; Tucker and Hitzman, 1994). The microbial prospecting method has been used to prioritize the drilling locations and to evaluate the hydrocarbon prospects of an area (Pareja, 1994), thus reducing risks and achieving higher success ratio in petroleum exploration.

4.1. Microbial oxidation of ethane, propane and n-butane

The microbial oxidation of ethane, propane and n-butane proceeds stepwise to alcohols, then to aldehydes and finally to acetate, which can be assimilated into cell material (Figure 2). The number of species of bacteria capable of using such hydrocarbons in this process increases with the expansion of the chain length of the alkanes. The degradation of the alkanes occurs by terminal oxidation by means of mono-oxygenase and by β-oxidation to acetyl-CoA, which is the initial substance in several biochemical reactions (Atlas, 1984).

5. Halo and apical anomalies

Distinct and definite petroleum gas seepage could be readily identified by geochemical means such as gas chromatographic analysis. There is a possibility that chemically detectable petroleum gases will be absent in the soil receiving micro seepages, due to microbial oxidation. Idealized model of bacterial effect on adsorbed soil gaseous hydrocarbons showed negative signal (Halo anomaly) for adsorbed soil gas and positive signal (Apical anomaly) for microbial activity. (Richers, 1985; Tedesco, 1995; Schumacher, 1996). The 'halo' theory may be reconciled on the basis of microbiology. It is observed that bacteria present over the petroleum accumulation would consume the petroleum gases adsorbed to the soil surface in this higher concentration zone thus, markedly decreasing the concentration of gas directly over soil receiving micro seepages, but the bacteria would not be stimulated to utilize the lower 'edge' concentrations. In soil-gas analysis, the edge or halo concentrations would show little difference from soil gas over the soil receiving microseepages. However, in soil analysis (gas desorption) samples from the halo region would actually have a higher concentration of hydrocarbons (Rosaire et al, 1940) since bacteria have not been stimulated to develop on hydrocarbons adsorbed to soil in this lower emanation area. The consumption of vertical migrating light hydrocarbons by bacteria will result in varying degrees of depletion of the hydrocarbons in the free soil gas and in hydrocarbons adsorbed soil gas (Richers 1985, Klusman 1993). The light hydrocarbons micro seepage is preferentially consumed over the area of highest seepage, resulting in a high rate of bacterial activity. The bacteria present over the petroleum accumulation would consume petroleum gases, whereas the edges will show high concentration of gases as not being utilized by bacteria where the microbial activity is very low. Thus geochemical technique of quantitative or qualitative adsorbed soil gas analysis may have limitations in place where high soil microbial activity exists resulting in nearly complete utilization of soil gases by microbes. The light hydrocarbon distribution will show a low signal directly over the source, resulting in a halo anomaly. The ratio of bacterial activity to hydrocarbon concentration will then exhibit an apical anomaly. The microbial indicators are therefore target specific, associated directly over the oil pool, where microbes flourish utilizing upcoming hydrocarbon gases. The Halo anomaly theory for hydrocarbon exploration is reconciled on the basis of microbiology, and has significant importance in hydrocarbon exploration (Horvitz, 1981; Jillman, 1987; Baum, 1994).

6. Sample collection method

The geo-microbial surveys of prospecting involve collection of suitable samples from sub soil horizon and detection of specific micro flora in the samples. Sampling is important since the validity of the test results depend largely upon the manner in which the samples are taken. The samples are collected using a hollow metal pipe by manual hammering. The soil samples of about 100 gm each were collected in pre-sterilized whirl-pack bags under aseptic conditions from a depth of about 0.5 to 1m (Davis 1967), and preferably on a grid pattern with a spacing of 200m. The samples were immediately stored at 2 to 4°C and subsequently transported to the laboratory and are stored cryogenic conditions till analysis. Soil samples should not be stored for long time, and samples should be analyzed as early as possible after collection.. Sampling should not be done in disturbed or excavated areas, soils contaminated with hydrocarbons, chemicals or animal wastes, swamps and areas under water shed. During collection of soil samples, rocks, coarse materials, plant residues, and animal debris have to be excluded.

7. Isolation of hydrocarbon oxidizing bacteria

The specific bacterial populations are measured by standard microbiological screening techniques for hydrocarbon-indicating microorganisms. A selective growth media is used which cultures only microorganisms capable of utilizing light hydrocarbons. Various methods such as Bacterial pellicle formation, Gas uptake, soil dilution and plating, soil sprinkled plating, clumped soil plating and Radio autography have been used by various researchers to determine the quantitative abundance of hydrocarbon oxidizing bacteria or oxidation of gaseous hydrocarbons. However gas uptake and soil dilution and plating methods have been widely used. Other techniques can determine the presence of living bacterial cells in soil samples. The determination of the number of colony-forming units (CFUs) in solid feeding media and the most-probable-number (MPN) procedure in nutrient solutions can be applied.

Isolation and enumeration of methane oxidizing bacteria and (C2+) ethane/propane/butane oxidizing bacteria for each sample is usually carried out by Standard Plate Count (SPC) method. 1 gm of soil sample was suspended in 9 ml of pre-sterilized water for the preparation of decimal dilutions (10^{-1} to 10^{-5}). A 0.1 ml aliquot of each dilution was placed on to Mineral Salts Medium (MSM) petri plates containing 1.0 g of $MgSO_4.7H_20$, 0.7 g of K_2HPO_4, 0.54 g of KH_2PO_4, 0.5 g of NH_4Cl, 0.2 g of $CaCl_2.2H_2O$, 4.0 mg of $FeSO_4.7H_2O$, 0.3 mg of H_3BO_4, 0.2 mg of $CoCl_2.6H_2O$, 0.1 mg of $ZnSO_4.7H_2O$, 0.06 mg of $Na_2MoO_4.2H_2O$, 0.03 mg of $MnCl_2.4H_2O$, 0.02 mg of $NiCl_2.6H_2O$, and 0.01 mg of $CuCl_2.2H_2O$ in 1000 mL of distilled water, at pH 7.0. These plates were placed in a glass desiccator, filled with the desired hydrocarbon gas (methane/ethane/propane with 99.99 % purity) and zero air (purified atmospheric gas devoid of hydrocarbons) in a ratio of (1:1). For isolation of methane oxidizing bacteria, the desiccator was filled with methane gas and zero air. Similarly, for isolation of ethane, propane and butane oxidizing bacteria, the desiccators were filled with either ethane/propane/butane gas with zero air respectively, whereas for isolation of n-pentane or n-hexane oxidizing bacteria, these plates were placed in glass desiccators containing air saturated with n-pentane or n-hexane vapor respectively (Rasheed, 2011). These desiccators were kept in bacteriological incubators at 35 ± 2°C for 10 days. After incubation, the developed bacterial colonies of methane, ethane, propane and butane oxidizing bacteria were manually counted using colony counter and reported in colony forming unit per gram (cfu/gm) of soil sample (Wittenbury et al., 1970; Rasheed et al. 2008).

8. Molecular biology techniques

Currently, molecular biology techniques has achieved great development in studies of soil samples. Development of molecular biology methods for microbial prospecting for oil and gas by applying culture independent techniques will improve the accuracy rate of microbial prospecting for oil and gas exploration (Fan Zhang et al., 2010). The most-probable-number (MPN) procedure has traditionally been applied to determine the numbers of colony forming units (CFUs) in soil samples. The real-time polymerase chain reaction (RT-PCR) is now being widely used to detect and quantify various target microorganisms without experimental cultivation (Dionisi et al. 2003; Skovhus et al. 2004; He et al. 2007). Molecular techniques related with 16S ribosomal DNA (RNA) have been proven effective as a basis for understanding the microbial diversity in environmental communities. The cloning and sequencing of 16S rDNA is sufficient for the identification of the microorganisms present in a given habitat and for the discovery of previously unknown diversity (Hugenholtz et al. 1998). These techniques were also applied to investigate microbial communities in the formation water of the produced water of oil fields (Kaster et al. 2009; Lysnes et al. 2009). Characterization of microbial communities involved in short-chain alkane metabolism, namely methane, ethane and propane, in soil samples from a petroliferous soils through clone libraries of the 16S rRNA gene of the Domains Bacteria and Archaea and the catabolic gene coding for the soluble di-iron monooxygenase (SDIMO) enzyme alpha subunit. Further studies on the occurrence and diversity of SDIMO genes in soil, as well as the improvement of primer sets to be applied in real-time PCR, are necessary in order to overcome the obstacle of the low abundance of catabolic genes in natural environments and enable their quantification from complex genetic backgrounds (Paula, et al., 2011). Immunological or DNA probes for gaseous hydrocarbon utilizing bacteria; Researchers are on for developing immunological probe or DNA probe which will rapidly identify the specific hydrocarbon oxidizers from the survey area. Immunological probes are based on the technique of preparing monoclonal antibodies as probes. In DNA probes, a selective small piece of DNA either from plasmid or from chromosomal segment serves as a probe. This segment is coded with the gene sequence, responsible for specific hydrocarbon gases oxidation. DNA hybridization with homologous strain gives the detection of hydrocarbon oxidizers. Detection of Biomarkers – Diploterol; The hapanoid diploterol is a known product of various aerobic bacteria that is particularly abundant in methanotrophic bacteria. The methanotrophs are in fact the source of diploterol is indicated by the compound light isotopic composition or around -60 per mil. Moreover 12C enriched hydrocarbon derivatives of diploterol were found in other studies of sediments in seep environments. This finding makes it a promising developmental technique for bio-prospecting of hydrocarbons. The above methods are the future thrust areas in bio-processing and will enhance the efficacy of the technique.

9. Plotting of bacterial anomaly

The results of hydrocarbon oxidizers population are plotted in terms of population density of aerial basis on the surveyed map using Arc GIS (Geographical Information System) or Golden Surfer Software's. A statistical approach has been followed and standard deviation value is taken as a background value for the demarcation of anomalous zones. The results of hydrocarbon oxidizing bacterial population are plotted on the surveyed map.

10. Evaluation of bacterial anomalaies

The samples showing bacterial population less than the background values indicate negative prospects, while the value above the standard deviation value gives the anomalies concentration of these gaseous hydrocarbon oxidizers. If the results of investigations are negatives; the completeness and the accuracy of the observations must be determined beforehand because the negative results may be due to various defects in the methods of field and laboratory work, such as insufficient sample taken at the location, inadequate reproducibility of bacterial cultures and keeping the samples for a long time. Bacterial anomalies detected in a soil survey may be classified as focal or continuous anomalies according to number of features depending upon a degree of localization of positive points. The anomalies are superimposed on prospect map or other geological or geophysical inputs for integration and interpretation. Classification of microbial anomalies according to the predominating types of indicator microorganism is also essential.

11. Case histories

We have presented some of the case histories of microbial surveys carried out in various sedimentary Basins of the Indian subcontinent by National Geophysical Research Institute (NGRI-CSIR), Hyderabad, India. A microbial survey was carried out in the established regions of the Kadi Kalol oil and gas fields of the Mehsana Block, Cambay basin, where the hydrocarbon utilizing bacteria ranged between 10^6 and 10^7 cfu/gm of soil (Table 1). In the other well-known areas, such as the Ponnamanda and Tatipaka gas fields of the Krishna Godavari basin, the hydrocarbon utilizing bacterial counts of these two areas were found to be 10^5 cfu/gm of soil, indicating the adaptation of microbes to utilize hydrocarbon seepage and possible presence of hydrocarbon deposits. The Oil and Natural Gas Commission (ONGC) of India has reported the presence of recoverable deposits of petroleum in these two areas. In the oil fields of Mehsana, and KG Basin, it is found that the number of hydrocarbon utilizing bacteria from a petroliferous area is in the range between 10^5 and 10^7 cfu/gm of soil. In the Jaisalmer gas fields the number of bacteria per gram of soil was always greater than 10^4 cfu/gm of soil. In the established oil and gas fields of Mehsana, Cambay basin, Jaisalmer basin and Krishna Godavari basin, soil samples showed bacterial growth $\geq 10^4$ account for 92%, 80% and 90% respectively.

The bacterial counts of these established oil and gas fields were ranged between 10^3 and 10^7cfu/gm, while in the exploratory area, the soil samples showed bacterial growth $\geq 10^4$, indicating that oil or gas exist in the latter area, which was eventually found to be correct after drilling operations (Rasheed et al, 2011) . The possibility of discovering oil or gas reservoirs by the microbiological method is emphasized by the fact that the count of hydrocarbon-oxidizing bacteria in soil or sediment samples ranged between 10^3 and 10^6 cfu/gm in soil/sediment receiving hydrocarbon micro-seepages, depending on the ecological conditions (Wagner et al, 2002).

Sampling area	Total no. of samples	Hydrocarbon utilizing bacteria (cfu/gm) of soil
Mehsana (oil/gas fields)	135	$10^6 - 10^7$
Jaisalmer (Gas fields)	100	$10^4 - 10^5$
Krishna Godavari Basin (Gas fields)	150	10^5
Shri Ganga Nagar Block, Rajasthan (Oil field)	150	10^5

Table 1. Hydrocarbon utilizing bacterial count of various established oil and gas fields in India.

11.1. Bikaner-Nagaur basin

A geo-microbial survey was carried out in the Shri Ganga Nagar Block of Bikaner-Nagaur basin, Rajasthan to investigate the prospects for hydrocarbon exploration. The propane oxidizing bacterial counts in the study area of the Shri Ganga Nagar Block were found to be ranged between 1.0×10^2 and 3.84×10^5 cfu/gm. The bacterial concentrations are plotted on the surveyed map using Arc GIS software. The bacterial concentration distribution maps of hydrocarbon utilizing bacteria show distinct anomalies in the studied area. The hydrocarbon oxidizing bacterial count ranged between 10^5 and 10^6 cfu/gm of soil, which is significant and thereby substantiate the seepage of lighter hydrocarbon accumulations from the subsurface petroleum reservoirs. The map of propane utilizing bacteria of the study area showed distinct microbial anomalies, which confirm the seepage of light hydrocarbons from the subsurface oil/gas reservoirs. The microbial results showed high propane oxidizing bacterial population in the studied area of the Bikaner Nagaur Basin, indicating positive prospects for hydrocarbon exploration. The GAIL (Gas Authority of India Limited) has reported the presence of recoverable deposits of petroleum in this area. (Figure 1).

11.2. Mehsana block, Cambay basin

Microbial prospecting studies were carried out in known petroliferous Mehsana Block of North Cambay Basin, India. A set of 135 sub-soil samples collected, were analyzed for indicator hydrocarbon oxidizing bacteria. The bacterial concentration map showed anomalous zones of propane oxidizing bacterial (Figs. 4).It is observed that the bacterial anomalies are found away from the oil and gas wells. The hydrocarbon microseepage is dependent upon pressurized reservoirs driving the light hydrocarbon microseepage upward. The pattern of reduced microbial counts adjacent to producing wells has been a commonly observed phenomenon for older producing fields. Over some well-drained gas reservoirs, the microbial values have been found to even be anomalously low. The phenomenon of apparent microseepage over the shutdown producing fields is thought to be due to a change in the drive mechanism controlling microseepage. When a well is brought into production, the drive mechanism changes from vertical, buoyancy driven force to horizontal gas streaming to the pressure sinks created around producing wells. When this occurs, microseepage greatly decreases or stops and microbial populations at the surface decline rapidly.

This change in drive mechanism and microbial population densities can be used to define reservoir drainage direction, radius, and heterogeneities around existing wells in producing fields (Tucker and Hitzman, 1994). In undrilled areas this phenomenon will not occur and there will be a direct relationship between high microbial populations, micro seepage, and potential reservoirs. Anomalous hydrocarbon microseeps are identified by observing bacterial population concentrations and distribution patterns within a survey area. There is a direct and positive relationship between the light hydrocarbon concentrations in the soils and these microbial populations. Surface contamination of produced oil and changing soil types do not affect these light hydrocarbons reflected in the microbial population distributions.

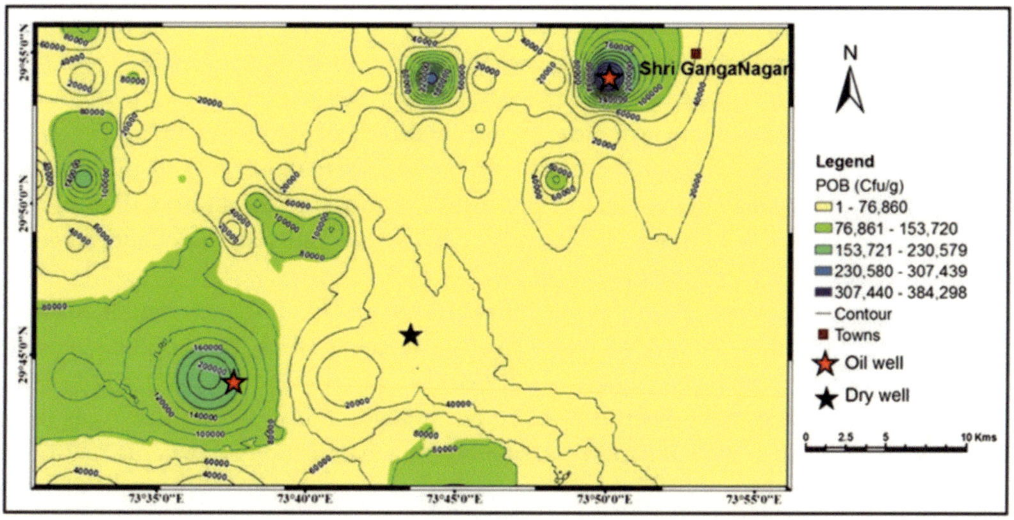

Figure 1. Results of microbial prospection studies in Shri Ganga Nagar Block, Rajasthan Basin, India.

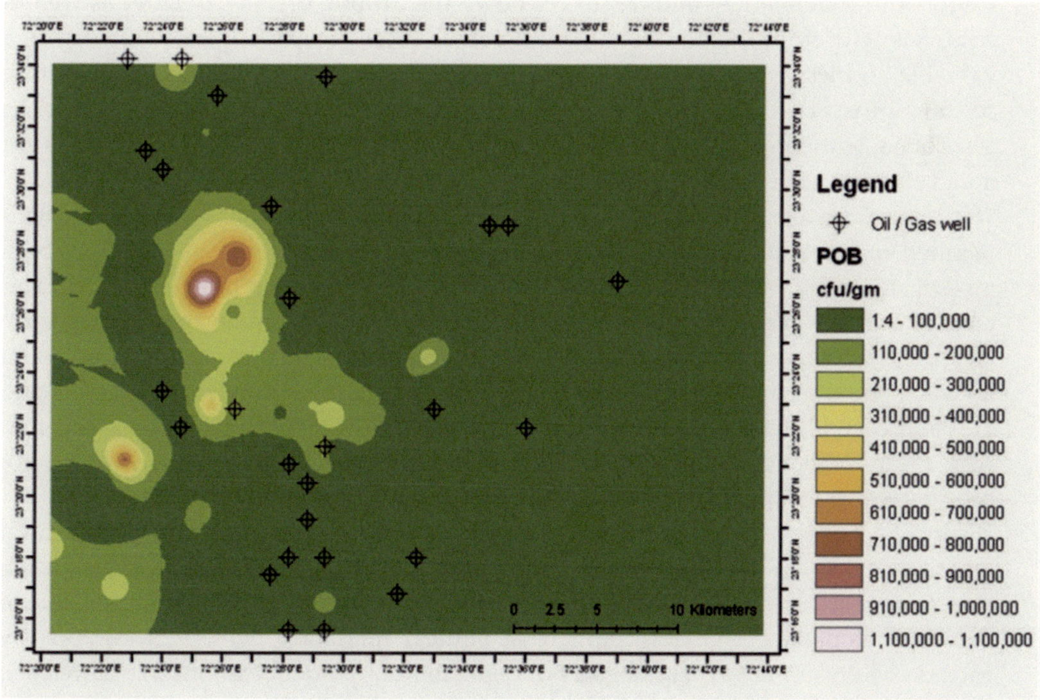

Figure 2. Map of Microbial survey using propane oxidizing bacteria (POB) in producing oil and gas field of Mehsana, Cambay Basin, Gujarat, India.

11.3. Advantages

i. Geo-Microbial prospecting method is well known in the realm of hydrocarbon exploration. Microbial prospecting method can be integrated with geological, geophysical and other surface hydrocarbon prospecting techniques. Microbiological methods have potential as a hydrocarbon exploration tool, development and extension of older fields. The microbial prospecting method is mainly used to prioritize the drilling locations and to evaluate the hydrocarbon prospects of an area (Pareja, 1994; Tucker and Hitzman, 1994 and Schumacher, 2000) thus reducing risks and achieving higher success ratio in petroleum exploration.

ii. One of the main advantages of the microbial prospecting method is, this method can be used for identification for prospective oil and gas in areas where no geophysical data is available, or where such investigation is difficult.

iii. Microbial anomalies have been proved to be reliable indicators of oil and gas in the subsurface (Pareja 1994). Hydrocarbon micro seepage detection adds value to 2-D and 3-D seismic by identifying those features that are charged with geomicrobial anomalies and sub-surface petroleum accumulations can be complex; on proper integration with geological and geophysical data, can contribute to the success of exploration and helps in risk reduction of dry wells. Since the drilling operations are costly, it is essential to use appropriate and efficient exploratory methods, either singly or in combination, in order to cut down the drilling cost of dry holes as well as wild cats with unprofitable recovery. Microbiological prospecting methodology is a valuable and less expensive value addition exploration tool to evaluate the valuable seismic prospects. This method can substantially reduce the exploration risks associated with trap integrity and hydrocarbon charge, especially in the hunt for much elusive subtle traps.

iv. The method can also be used independently and basically no geological or seismic data is required to carry out microbial prospecting surveys. In areas that have not yet been investigated geophysically, this technique can be applied as wildcat prospecting tool. This method can give principal evidence on the occurrence of hydrocarbon anomalies in large areas. The subsequent seismic and geological investigations can thus be, concentrated on favorable areas in those regions where structure data of the sub-surface already exists. This approach likewise does not require any knowledge of the position of the structures. As a result therefore, the seismic structure maps and microbial anomalies, which have been drawn up independently from one another, can be compared and contrasted.

v. Distinct and definite petroleum gas seepage could be readily identified by geochemical means such as gas chromatographic analysis. There is a possibility that chemically detectable petroleum gases will be deficient in the soil receiving micro-seepages, due to microbial oxidation. The consumption of vertical migrating light hydrocarbons by bacteria will result in varying degrees of depletion of the hydrocarbons in the free soil gas and in hydrocarbons adsorbed soil gas (Richers 1985, Klusman 1993). The light hydrocarbons micro seepage is preferentially consumed over the area of highest seepage, resulting in a high rate of bacterial activity. The bacteria present over the petroleum accumulation would consume petroleum gases, and the edges will show high concentration of gases as not being utilized by bacteria where the microbial activity is very low. Thus geochemical technique of quantitative or qualitative adsorbed soil gas analysis may have limitations in place where high soil microbial activity exists resulting in partial or complete utilization of soil gases by microbes. The light hydrocarbon distribution will show a low directly over the source, resulting in a halo anomaly. The ratio of bacterial activity to hydrocarbon concentration will then exhibit an apical anomaly. The microbial indicators are therefore target specific, associated directly over the oil pool microbes flourish utilizing upcoming hydrocarbon gases. The Halo anomaly theory for hydrocarbon exploration is reconciled on the basis of microbiology, and has significant importance in hydrocarbon exploration (Horvitz, 1981; Jillman, 1987; Baum, 1994).

vi. Indeed, it is good supplementary tool for hydrocarbon prospecting and on proper integration with geological and geophysical data, can contribute to the success of exploration and helps in risk reduction of dry wells.

vii. According to the authors, microbial prospecting studies have to be taken up with the adsorbed soil gas analysis. As distinct and definite petroleum gas seepage could be readily identified by geochemical means such as Adsorbed soil gas analysis. There is a possibility that chemically detectable petroleum gases will be deficient in the soil receiving micro seepages, due to microbial oxidation. Thus geochemical technique of quantitative or qualitative adsorbed soil gas analysis may have limitations, where high soil microbial activity exists resulting in partially or nearly complete utilization of soil gases by microbes. Therefore, it is crucial to perform microbial prospecting studies along with the adsorbed soil gas analysis.

Author details

M.A. Rasheed*, D.J. Patil and A.M. Dayal
Microbiology laboratory, Petroleum Geochemistry Group, National Geophysical Research Institute, Council for Scientific and Industrial Research (CSIR), Hyderabad, India

12. References

Atlas, R. M., 1984. Petroleum microbiology: New York, Macmillan Company, 692 p.

Baum, M. G. 1994. Applied geochemical surface investigations as a supplemental exploration tool—Case histories: Oil and Gas Erdgas Kohle, v. 110, n. 1, p. 9–15.

Beghtel, F.W., Hitzman, D.O., and Sundberg, K.R., 1987. Microbial Oil Survey Technique (MOST) evaluation of new field wildcat wells in Kansas. Association of Petroleum Geochemical Explorationists Bulletin, 3, 1-14.

Blau, L. W., 1942, Process for locating valuable subterranean deposits: U. S. Patent 2,269,889.

* Corresponding Author

Bokova, E.N., Kuznetsova, V.A. and Kutznetsova, S.I., 1947. Oxidation of gaseous hydrocarbons by bacteria as a basis of microbiological prospecting for petroleum, Dokl. Akad. Nauk S.S.S.R., 56: 755-757.

Brisbane, P.G., and Ladd, J.N, 1965. The role of microorganisms in petroleum exploration, Annual Review Microbiology, vol.19, PP.351-364.

Davis, J. B., 1956, Microbial decomposition of hydrocarbons: Industrial and Engineering Chemistry, v. 48, no. 9, p. 1444–1448.

Davis, J. B., 1967, Petroleum microbiology: Elsevier Publishing Company, p. 197–245.

Dionisi, H.M., Harms, G., Layton, A.C., Gregory, I.R., Parker, J., Hawkins, S.A., Robinson, K.G., and Sayler, G.S. (2003). Power analysis for real time PCR quantification

of genes in activated sludge and analysis of the variability introduced by DNA extraction. Appl. Environ. Microbiol. 69:6597–6604.

Fan Zhang., Yuehui She., Yong Zheng., Zhifeng Zhou., Shuqiong Kong and Dujie Hou (2010). Molecular biologic techniques applied to the microbial prospecting of oil and gas in the Ban 876 gas and oil field in China. Appl. Microbiol Biotechnol, 86:1183–1194. DOI 10.1007/s00253-009-2426-5.

Hanson, R. S. and Hanson, T. E., 1996, Methanotrophic bacteria. Microbiol.Rev., 60, 439–471.

Hanson, R. S. and Wattenberg, E. V. 1991, Ecology of methylotrophic bacteria, in I. Goldberg and J. S. Rokem, eds., Biology of methylotrophs: London, Butterworths Publications, p. 325–349.

He, J.Z., Shen, J.P., Zhang, L.M., Zhu, Y.G., Zheng, Y.M., Xu, M.G., Di, H.J. (2007). Quantitative analyses of the abundance and composition of ammonia-oxidizing bacteria and ammonia-oxidizing archaea of a Chinese upland red soil under long-term fertilization practices. Environ. Microbiol. 9:2364–2374.

Hitzman, D. C., J. D. Tucker, and P. D. Heppard, 1994, Offshore Trinidad survey identifies hydrocarbon microseepage: 26th Annual Offshore Technology Conference, OTC 7378, Houston, Texas.

Hitzman, D.C., 1999. 3-D seismic and hydrocarbon micro seepage: Exploration adventures in Osage County, Oklahoma, DOE & OGS Silurian,Devonian & Mississippian Geology and Petroleum in the Southern Mid Continent Workshop, Norman, O.K.

Horvitz, L., 1939. On geochemical prospecting-I, Geophysics, Vol. 4, PP. 210-228.

Horvitz, L., 1981, Hydrocarbon prospecting after forty years. In Unconventional Methods in Exploration for Petroleum and Natural Gas 2 (ed. Gottleib, B. M.), Southern Methodist University Press, Texas, pp. 93–95.

Hugenholtz, P., Goebel B.M., and Pace, N.R. (1998) Impact of cultured independent studies on emerging phylogenetic view of bacteria diversity. J. Bacteriol. 180:4765–4774.

Jain, A.K., 1991. Microbial Prospecting-A Promising Technique for Hydrocarbon Exploration. In Proceedings of International Conference on Science and Technology, Nov.91, New Delhi. India.

Jillman, N., 1987. Surface geochemistry enigmas. Oil Gas J., 9, 87–89.

Kartsev, A.A., Tabasaranskii, Z.A., Subbota, M.I., and Mogilevskii, G.A., 1959, Geochemical methods of prospecting and exploration for petroleum and natural gas: Berkeley, University of California Press, 349.

Kaster, K.M., Bonaunet, K., Berland H., Kjeilen-Eilertsen, G., Brakstad, O.G. (2009) Characterisation of cultured-independent and –dependent microbial communities in a high-temperature offshore chalk petroleum reservoir. Antonie Van Leeuwenhoek 96:423–439.

Klusman, R. W., 1993, Soil gas and related methods for natural resource exploration: New York, John Wiley and Sons, 483

Laubmeyer, G., 1933, A new geophysical prospecting method, especially for deposits of hydrocarbons: Petroleum London, v. 29, p. 14.

Leadbetter, E.R. and Foster, J.W. 1958, studies on some methane utilizing bacteria. Archives Microbiology. 30, 91-118.

Lopez, P.J., Tucker, J.D. and Hitzman, D.C., 1993, Hydrocarbon microseepage signatures of seismic structures identified by microbial surveys, sub-Andean region, Bolivia, AAPG Annual convection Program p.162.

Lysnes, K., Bodtker, G., Torsvik, T., Eva., Bjornestad, E.S. (2009). Microbial response to reinjection of produced water in an oil reservoir. Appl Environ Microbiol 83:1143–1157.

Miller, G. H., 1976. Microbial survey helps to evaluate oil and gas. Oil and Gas Journal, 4, 192.

Mogilevskii, G.A., 1959, Geochemical methods of prospecting and exploration for petroleum and natural gas: Berkeley, University of California Press, 349 p.

Mogilewskii, G. A., 1938, Microbiological investigations in connecting with gas surveying: Razvedka Nedr, v. 8, p. 59–68.

Mogilewskii, G. A., 1940, The bacterial method of prospecting for oil and natural gases: Razvedka Nedr, v.12, p. 32–43.

Paula B Miqueletto, Fernando D Andreote, Armando CF Dias, Justo C Ferreira, Eugênio V dos Santos Neto and Valéria M de Oliveira. Cultivation-independent methods applied to the microbial prospection of oil and gas in soil from a sedimentary basin in Brazil. Miqueletto et al. AMB Express 2011, 1:35 http://www.amb-express.com/content/1/1/35

Pareja, L., 1994, Combined microbial, seismic surveys predict oil and gas occurences in Bolivia. Oil Gas J., 24, 68–70.

Perry, J. J., and William, S., 1968. Oxidation of hydrocarbons by microorganisms isolated from soil. Canadian Journal Microbiology, 14, 403–407.

Rasheed, M. A. Veena Prasanna,M. Satish Kumar,T. Patil, D.J. and Dayal, A. M. 2008, Geo-microbial prospecting method for hydrocarbon exploration in Vengannapalli Village, Cuddapah Basin, India, Current Science, Vol.95, no.3 p. 361-366.

Rasheed M.A., Lakshmi M., and Dayal A.M (2011). Bacteria as indicators for finding oil and gas reservoirs: A Case Study of Bikaner-Nagaur Basin, Rajasthan, India. Petroleum Science Journal. Volume 8, Number 3, 264-268.

Richers, D.M., Reed, R.J., Horstman, K.C., Michels, G.D., Baker, R.N., Lundell, L., and Marrs, R.W., 1982. Landsat and soil-gas geochemical study of Patrick Draw Oil Field, Sweetwater County, Wyoming: American Association of Petroleum Geologists Bulletin, v. 66, p. 903–922.

Richers, D.M. 1985. Some methods of incorporating remote sensing and surface prospecting with hydrocarbon exploration. In Surface and near surface Geochemical methods in Petroleum Explorationists, special Publication no.1, Denver, CO, pp. C1-C93.

Rosaire, E. E., 1939, The handbook of geochemical prospecting: Houston, Texas, Gulf Publishing Co., 61 p.

Schumacher, D., 1996. Hydrocarbon-induced alteration of soils and sediments, in Schumacher, D., and Abrams, M.A., eds., Hydrocarbon migration and its near-surface expression: American Association of Petroleum Geologists Memoir 66, p. 71–89.

Schumacher, D., Hitzman, D.C., Tucker, J., Rountree, B., 1997. Applying high-resolution surface geochemistry to assess reservoir compartmentalization and monitor hydrocarbon drainage. In: Kruizenga, R.J., Downey, M.W. (Eds.), Applications of Emerging Technologies:

Sealy, J. R., 1974a. A geomicrobial method of prospecting for oil. Oil and Gas Journal, 8, 142–46.

Sealy, J. R., 1974b. A geomicrobial method of prospecting for oil. Oil and Gas Journal, 15, 98–102.

Skovhus, T.L., Ramsing, N.B., Holmstrom, C., Kjelleberg, S., Dahllof, I. (2004). Real-time quantitative PCR for assessment of abundance of pseudo alteromonas species in marine samples. Appl. Environ. Microbiol. 70:2373–2382.

Taggart, M. S., 1941, Oil prospecting method: U. S. Patent 2,234,637.

Tedesco, S.A., 1995, Surface Geochemistry in Petroleum Exploration: New York, Chapman and Hall, Inc., 1 -206 p.

Tucker, J., and D. C. Hitzman, 1994, Detailed microbial surveys help improve reservoir characterization: Oil and Gas Journal, v. 92, no. 23, p. 65–68.

Updegraff, D. M., and H. H. Chase, 1954, Microbiological prospecting method: U. S. Patent 2,861,921.

Vestal, J.R., and Perry, J.J., 1971. Effect of substrate on the lipids of the hydrocarbon utilizing Mycobacterium vaccae. Canadian Journal of Microbiology, 17: 445-449.

Wagner, M., Wagner, M., Piske, J. and Smit, R., 2002, Case histories of microbial prospection for oil and gas. AAPG Studies in Geology 48 and SEG Geophysical References Series, vol. 11, pp. 453–479.

Whittenbury, R., K. C. Phillips, and J. G. Wilkinson, 1970, Enrichment, isolation and some properties of methane utilizing bacteria: Journal of General Microbiology, v. 61, p. 205–218.